SpringerBriefs in Philosophy

SpringerBriefs present concise summaries of cutting-edge research and practical applications across a wide spectrum of fields. Featuring compact volumes of 50 to 125 pages, the series covers a range of content from professional to academic. Typical topics might include:

- A timely report of state-of-the art analytical techniques
- A bridge between new research results, as published in journal articles, and a contextual literature review
- A snapshot of a hot or emerging topic
- An in-depth case study or clinical example
- A presentation of core concepts that students must understand in order to make independent contributions

SpringerBriefs in Philosophy cover a broad range of philosophical fields including: Philosophy of Science, Logic, Non-Western Thinking and Western Philosophy. We also consider biographies, full or partial, of key thinkers and pioneers.

SpringerBriefs are characterized by fast, global electronic dissemination, standard publishing contracts, standardized manuscript preparation and formatting guidelines, and expedited production schedules. Both solicited and unsolicited manuscripts are considered for publication in the SpringerBriefs in Philosophy series. Potential authors are warmly invited to complete and submit the Briefs Author Proposal form. All projects will be submitted to editorial review by external advisors.

SpringerBriefs are characterized by expedited production schedules with the aim for publication 8 to 12 weeks after acceptance and fast, global electronic dissemination through our online platform SpringerLink. The standard concise author contracts guarantee that

- an individual ISBN is assigned to each manuscript
- each manuscript is copyrighted in the name of the author
- the author retains the right to post the pre-publication version on his/her website or that of his/her institution.

More information about this series at http://www.springer.com/series/10082

Polona Tratnik

Conquest of Body

Biopower with Biotechnology

 Springer

Polona Tratnik
Institutum Studiorum Humanitatis, Faculty
  for Postgraduate Studies in Humanities
Alma Mater Europaea
Ljubljana
Slovenia

ISSN 2211-4548   ISSN 2211-4556   (electronic)
SpringerBriefs in Philosophy
ISBN 978-3-319-57323-6   ISBN 978-3-319-57324-3   (eBook)
DOI 10.1007/978-3-319-57324-3

Library of Congress Control Number: 2017938142

Printed on acid-free paper

This Springer imprint is published by Springer Nature
The registered company is Springer International Publishing AG
The registered company address is: Gewerbestrasse 11, 6330 Cham, Switzerland

To Gašper

Acknowledgements

I pay tribute to Siegfried Zielinski, the Former Director of the Vilém Flusser Archive in Berlin and Founder of the media archeology discipline, now the Director of the Zentrum für Kunst und Medien in Karlsruhe, for his attention to the concept of projection originating in Flusser and to the related concept *perspicere*, which in several aspects presents an opposition to *proicere*. It was November 2010 in Milan at a conference on new media art when we discussed the philosophy of media. At the time, his short study on the genealogy of projection was forthcoming,[1] and I just became an instructor of a university course on media culture which I had turned into a class on media philosophy. Thus, I was extremely attracted by the issues of media as mediation, the functioning of representation in photography and other visual media, predicates in writing and speech, the linguistic turn, the pictorial turn, processes of communication, performativity, etc. At the same time, my book on mediation, performativity, and rhizomaticity was coming out. In it, I refer mostly to photography and biotechnological art as both aiming to make live beyond the body.[2] Both then and now, I have been strongly influenced by poststructuralist thought.

Interestingly enough, the interest in media theory and philosophy has risen over the past decade: There have been numerous McLuhan media centers established worldwide (2011 marked the centenary of McLuhan's birth), and new media art centers and festivals have paid exhaustive attention to media issues, particularly emphasizing Flusser's and McLuhan's respective contributions (2011 Transmediale addressed McLuhan versus. Flusser versus. McLuhan versus. Flusser) and thus recognizing the two as the founding fathers of media theory and philosophy. Last but not least, it is an important fact, not only for the English-speaking world but also globally (since English has become the new Latin and has spread out even

---

[1] Siegfried Zielinski, *Entwerfen und Entbergen. Aspekte einer Genealogie der Projektion* (Köln: Walther König Verlag, 2010).

[2] Polona Tratnik, *Hacer la presentia. Fotografía, arte y (bio)tecnología* (Ciudad de Mexico: Herder, 2013).

more than Latin in the past both vertically [it is not spoken only by scholars in academic spheres, but has become the language for all occasions and everyone] and horizontally [now telecommunication is working for Flusser]), that the English translation of Flusser's most influential work so far (*Towards a Philosophy of Photography*, written in 1983) did not appear before 2000. The first translation of his books into English appeared one year earlier (*The Shape of Things*), while more translations have only recently appeared (Vampyroteuthis Infernalis, *Does Writing Have a Future?*, and *Into the Universe of Technical Images* were all published in 2011, and Flusser's *Writings* came out in 2004), meaning the world has only just now started to acknowledge the importance of his theories.

When I was preparing a paper for a conference on Marshall McLuhan and Vilém Flusser's Aesthetics and Communication Theories Revisited at the Video Pool Media Arts Centre in McLuhan's home town of Winnipeg in June 2012, I focused on Flusser's theory of media and related it to biotechnology. In particular, I discussed *proicere* and *perspicere* in regard to body imagery, the exploration of the body, and the relation between gaining knowledge and engineering. The reader finds an extended study of this subject in the first chapter. The majority of the book was written during my stay at the University of California Santa Cruz as a Fulbright Visiting Scholar from February to September 2012. I am thankful to the Council for International Exchange of Scholars for this support, which actually presented the very material ground for this book. I am also grateful to my colleagues from this University for various kinds of support during my stay in Santa Cruz: first of all to my sponsor Tyrus Miller; then to the Department of History of Art and Visual Culture that affiliated me, particularly to Martin Berger. The Art Department was also very cooperative, and I am particularly happy to have met and collaborated with Elizabeth Stephens. For the referential support in getting the grant, I thank Aleš Erjavec, Miško Šuvaković, and Jurij Krpan. The research was partly conducted within the projects in which I have been taking part at the Science and Research Center of the University of Primorska in Slovenia, particularly the investigation of biotechnological issues and aspects of biopower. I am grateful to Nadja Furlan, Lenart Škof, and Darko Darovec for enabling this research.

A shorter version of the final chapter has been published in *Annales. Series Historia et Sociologia* (vol. 22, 2012, nr. 2). I included parts of the chapter "Solution of Life" in the article with another focus as presented here—I discussed my own art practice (*Technoetic Arts: A Journal of Speculative Research*, vol. 13, 2015, nr. 1 & 2). The early version of the second chapter concerning art was published in *Maska* (summer 2011, vol. 26, nr. 139/140) and was for that occasion translated from Slovenian to English by Eva Erjavec. For this book, I revised and extended that study. I thank the editors for allowing the reprint of the text here. For his precise and attentive language revision of this book, I express my gratitude to Erik Bachman. For helping me with the final language revisions, I am thankful to Goran Gumze.

Some theories explained in this book have already been presented at various conferences since 2011. I am thankful to the organizers of these meetings for stimulating inspiring discussions in these related issues, particularly to Zorica

Ivanović and the Faculty of Arts Belgrade for conducting a debate on the body, biotechnology, and power and to the Society for Phenomenology and Media for devoting attention to media philosophy. I am much obliged for the invaluable aid in getting the necessary knowledge and practical experience in the field of biotechnology that was provided to me by my biotechnological collaborators: Tamara Lah Turnšek, Miomir Knežević, Helena Motaln, Primož Rožman, Marko Strbad, Ajda Marič, and other collaborators from the National Institute of Biology Slovenia, the Blood Transfusion Center Slovenia, Biobanka, and Educell. At this occasion, I would want to express my deep gratefulness to Ludvik Toplak for trusting me the responsible position of the Dean of the Faculty for Humanities Alma Mater Europaea—Instiutum Studiorum Humanitatis. This gave me the opportunity to significantly contribute to the field of humanities and also to accomplish edition of this book. Last but not least, I am thankful to my students from Slovenia, the USA, China, and elsewhere, from whom I have always learned a lot. Certainly, this book would never have been written without the support of my family. I give the warmest hug to my most beloved ones.

Ljubljana, Slovenia   Polona Tratnik
October 2016

Contents

Introduction

> The relation of knowledge to power, as we were able to observe on the occasion of the conquest, is not contingent but constitutive.[3]
>
> Tzvetan Todorov

This book represents a discussion on the actual phenomenon of biotechnology and how it affects the body. Several related issues concerning human body and life are analyzed, particularly the issues of visualization, mediation, and epistemology. We defend the thesis that the exploration of human body is ultimately aimed at the conquest of it. In his study of mapping, Stephen S. Hall discusses the relation between scientists and conquistadors that Tzvetan Todorov recognized as constitutive of this conquest, and he poses a question: "Does the curiosity that drives exploration and discovery invariably lead to some sort of imperialism?"[4] In a manner similar to how the territories were discovered, explored, comprehended, mapped, conquered, and utilized, the knowledge we have been gaining about the body is constitutive of the power over it. Knowledge works for the power over life —that is to say for the politics of an individual body and the politics of life of a population. We are currently witnessing extensions and reaches of biopower as the power over life that were not possible before, primarily due to the engineering applications and prospects of biotechnology.

Security and ethical issues have to be taken care of regularly and accurately, based on complex and profound discussions and estimation of risks as a basis for actual risk management. Politics aims at regulating and limiting the reach of technology, still, some authors believe that if something is possible to do, it will be done.[5] For our part, we would put it this way: Which development is to be fostered,

---

[3]Tzvetan Todorov, *The Conquest of America*, trans. Richard Howard (New York: Harper & Row, Publishers, Inc., 1984) p. 181.

[4]Stephen S. Hall, *Mapping the Next Millennium. How Computer Driven Cartography Is Revolutionizing the Face of Science* (New York: Vintage Books, 1993), p. 383.

[5]For such an account see: Kevin Kelly, *What Technology Wants* (New York: Viking, 2010).

and which regulations will be implemented depends upon the fact of whose interests are at stake and who is in the position of power. To avoid the naïve and simple subjection to the politics of power, everyone needs to feel the urge to take part in the debate about the goals, risks, and interests that will result in decision making. In order to do that, the humanity in general needs to be better educated and informed. We agree with Manuela Monti and Carlo Alberto Redi, who call for public responsibility about biotechnological engagement.[6]

Furthermore, it is important that we raise awareness about how the so-called side effects of the development of technology come into existence with the attainment of a favored goal—in other words, how an effect that is generally regarded as an advantage welcomed by everybody nevertheless triggers great changes in seemingly unrelated spheres, e.g., in the economic infrastructure of society. It is easier to close one's eyes to the far-reaching effects of the development of technology than it is to take responsibility for one's part in increasing social inequality. For example, people would tend to agree that we should foster the development of biotechnology for medical purposes because it offers the hope of someday curing cancer. However, if the treatment is very expensive (and it will surely have its price), only the rich will be able to afford it. This fact then draws a political cartography of power by enlarging economic differences between people and nations. The techno-medical issue becomes a class problem or a problem of the global arrangement of political power. Micro-body interventions have macro-effects on economic and political situations. Moreover, there is a recurring dilemma here: Shall we salute biotechnological developments that will prolong the lifetime of individuals and of population as a whole—a prominent paradigm supporting this is the paradigm of the regenerative body, which we discuss in the final chapter—while on the other hand, the globe continues to be overpopulated. At the same time, some of us are impatient about the exploration of space, which we expect to eventually deliver the good news that we are finally able to migrate to another planet and

---

[6]"The scientific community needs political decision-makers to develop valid policy guidelines upon which to base the management of the wide-ranging issues generated by the biotechnological revolution and by SC [stem cell] research. Of extreme importance and urgency is the governance of biotechnology research, a process to which all citizens should be able to contribute. To do so, citizens will have to expand their understanding of the intrinsic opportunities and limitations inherent in biotechnology, and particularly in SC biology and techniques. Ideally, each individual should be able to constantly re-delineate the boundaries of his relationship with the world, and thus to stay abreast of the changes wrought in boundaries, as well as in the deeper meaning of life and its forms, by our era's constant advances in knowledge: the biopolitics of the body and its transformations, birth, end-of-life decisions, biomedical experimentation and control of personal decisions, are all at issue. Only through the development of reflective attitudes will citizens avoid the *easy routes* in debate, e.g. the opinion that the research described here 'will lead to technology taking control over humans' would be (horribly) *easy*. A perfect example of the impediment of the unreflective, *easy route* thinking is to be found in the story of cloning." Manuela Monti and Carlo Alberto Redi, "Stem Cells," in: Alfonso Barbarisi (ed.), *Biotechnology in Surgery. Updates in Surgery*, Vol. 0 (Milan: Springer Verlag Italy, 2011), p. 144.

colonize it, thus saving ourselves and our species from the fallout of greater environmental changes we may not be able to survive given the modes of life and industrial practices to which we currently adhere. We want to get out of here because the world is too small after all—it has been explored, cultivated, utilized, crowded, and restricted.

Western civilization is a conquering civilization per se. Therefore, modern man wants to relive the story of conquest again and again. And here is a conquest to be found right where we all already are: The world may be getting smaller and smaller, but the body is getting vaster and greater. It is actually becoming enormous, particularly when one aims to master it. The more open it gets and the more it is pierced into, the more there is to explore. According to the Renaissance episteme, man only had one body bounded by skin, a functional active body with rather simple mechanics. However, in accordance with today's episteme, there is a multitude of microorganisms spreading into this milieu and vice versa, a condensation of life organized in a very complex way. We no longer possess a solid static corpus, there is no firm boundary; one sees through the opaque body membranes and slips into a detail, learns about a cell and genome, the program of life. Here, we are faced with numerous questions that remain unanswered: What is actually the importance of these microorganisms, how do they function and interact, how exactly is the mind embedded in the flesh, how does it work, how does the genome work, etc.? And then there are related questions that concern power, among them: How do we reach the nano-level, how are we to intervene and act in a purposive way at that level, and how ought we best develop successful body engineering? Furthermore, in the context of fear—some national politics strongly stimulate this emotion in the body of their populations, which justifies the strengthening of the repressive apparatus of the state. This means an increase of surveillance mechanisms capable of mapping every one of us. If the main function of a prison is to suppress the freedom of living, then this would make the world and our bodies an interlocked prison. In this book, we discuss the point where medicine and the prison meet: in biopower, in the suppression of one's power, in the conquering of the body and its subjection to the mechanisms of power. Furthermore, as demonstrated in another contemporary story involving the incursion of surveillance mechanisms into our daily lives (namely the one about the control of the World Wide Web), the information gathered here is a precious commodity for which there are great appetites, which means that "private" information is increasingly becoming an instrument for the accumulation of capital and consequently of power. To point at such connections and broader effects, particularly when actual development has not yet demonstrated its full ramifications, humanistic reflection is of extreme importance. In a certain sense, things simply do not exist if they are not enlightened and discussed.

***

In this book, we discuss the conquest of body. For its early explorers, the eye was an instrument of knowledge. Researchers have been an instrument of the conquest. We link the quest for knowledge to the quest for power. Accordingly, the conquest of body does not just mean gaining knowledge of the body, as well as of life, but it also means subjecting this knowledge to the aim of mastering the body, serving the politics of quality optimization, and healing it in order to disengage it from the threats of illnesses and death. Ultimately, the conquest of body means the power to intervene into life processes. Biotechnology is the technology that enables this intervention *par excellence*.

Chapter 1 is devoted to the discovery of the body, the construction of knowledge about the body, and the aspiration to interfere with its "natural" processes. Along with this investigation, the crucial epistemological issues are being discussed. The presentations or representations of the body, the processes of gaining knowledge about it and of gaining power over it, are comprehended as interrelated. Anatomical drawings of the human body were relevant for the conceptual organization of this subject (human body in general) and *De Humani Corporis Fabrica* (Vesalius 1543), which is considered the first scientific anatomy book, served the medical practice. Anatomy pierced into the body to see through the opaqueness of tissues and visual representations were based on the same logic of revealing what is hidden to the naked eye. In Renaissance, the regime of media transparency became commonly accepted. The walls opened with painted "windows" through which the eye could see various environments. The method supported planning, the drawing became a sketch for the future building, and visualizations were involved in the practice of engineering. In this chapter, visual media that represent human body are questioned as means for discovering or "insight" into the truth and discussed as related with the tendency to gain power over the body. There are three moments of visualization that are closely examined in the first subchapter—the principle of mimesis, the technique of perspective, and map making. In the second subchapter, the tendency to pierce deeper in the body is analyzed from aesthetic and epistemological aspects. With the birth of the clinic, anatomy has started to serve medicine and the gaze of the nineteenth-century anatomist became focal. The focus of the third subchapter is this moment, in which knowledge quite directly came in service of intervention. Yet, there was a delay of intervention into the body because dead bodies were observed in the nineteenth century. This fact prevented timely healing of the ill bodies. Technical imaging on the contrary enables observation of a living ill body. Enormous development of imagery technologies throughout the twentieth century therefore crucially strengthened the power over the body. In the fourth subchapter, gaining power over body is discussed in reference to biotechnology. The issue of godlike creating, which is discussed in the first subchapter as regards the principle of mimesis, is here re-examined in the case of synthetic biology.

Two terms are central to the discussion: *perspicere* and *proicere*. To be able to comprehend or gain insight (*perspicere*), we *project*. We project what is then to be acknowledged. What then does the quest for the truth mean in this regard? What is this conquest all about? These questions become seemingly even more actual with the intervention of biotechnology because now we manipulate matter according to our own program, we are engineering and designing. In other words, we are creating a world in which we want to live. At the same time, we strive to explore the universe in order to be able to migrate to other planets. Today, the body has become subjected to the technology of power—i.e., biopower—to such a degree that the power over life itself has been exercised not only over the population in general, but also on the fleshy, micro-, nano-, chemical, and visual scale of a singular body. Biotechnology does indeed give humankind a power over life that was not attainable before, though the intervention of biotechnology belongs to a quest that did not first begin in the biotech century. Moreover, it is a part of a quest that has not yet been completed. But this power is not omnipotent. We are able to intervene in certain ways, but this does not mean that we can intervene in every possible way, nor are we able to create a human being from scratch. Is there a grand narrative here about the attainment of a totalitarian power over life that one can detect in the aspiration to this ultimate goal to "create" life from scratch and which is in turn detectable in the "development" of knowledge about life and the technology designed to manipulate it?

In Chap. 2, genetic prints are examined as representations in relation to what they represent. In science and repressive apparatus of the state, body prints are considered to be authentic indexical signs, while in some recent art projects, the genetic language and the issue of identification are challenged. The relations between the signifying structures and the body are analyzed in this chapter. Collecting body prints and the issue of identification are comprehended as strategies for biopower to be profoundly exercised over individual bodies and over populations (biopolitics).

A knowledge of life is the ground for defining life and death. These definitions ensure the political power over the body. With the development of technical devices that enable to see more and more and to "touch" the scales which the naked eye cannot see, the body is becoming an extensive milieu and knowledge of life changes, as we argue in Chap. 3. Today, life can get dissolved, dispersed, diluted, or delayed. Precisely, the ascertainment of the interrelations of the processes of living and dying affects the comprehension of body, and this increases the biotechnological power of intervention.

In the last chapter, we discuss the paradigm of the regenerative body, which is particularly interesting since it has discovered the quality that enables us not only to distinguish life from mechanics, but also to intervene into life processes in order to "improve" or "rescue" the body from dying or aging. This is ensured by the quality of regeneration. Regenerative body generates an ultimate dream of the conquest of body: an immortal active life of a body in constant process of vitalization, with which the process of mortification is defeated once and for all.

***

We are conscious of the modern hermeneutical acknowledgement about the intertwinement of the text and its interpreter. We do investigate the subject here, and we offer interpretations of it. Perhaps today the reader is aware that all concepts—including the ones discussed here—are constructs created by culture. They change not only with a given culture, but also with the positions we adopt toward them within that culture. For this reason, they change according to how we comprehend them. After all it is we who *project*.

Reference

A. Vesalius, *De Humani Corporis Fabrica* (1543). Available at: National Library of Medicine, http://archive.nlm.nih.gov/proj/ttp/flash/vesalius/vesalius.html. 6-1- 2012

Chapter 1
Conquest of Body: Mapping, Knowing, Mastering

Abstract The first chapter is devoted to the discovery of the body, the construction of knowledge about the body and the aspiration to interfere with its "natural" processes. Along with this investigation, the crucial epistemological issues are being discussed. The presentations or representations of the body, the processes of gaining knowledge about it and of gaining power over it are comprehended as interrelated. Anatomical drawings of the human body were relevant for the conceptual organization of this subject (human body in general) and *De Humani Corporis Fabrica* (Vesalius 1543), which is considered the first scientific anatomy book, served the medical practice. Anatomy pierced into the body to see through the opaqueness of tissues and visual representations were based on the same logic of revealing what is hidden to the naked eye. In Renaissance the regime of media transparency became commonly accepted. The walls opened with painted "windows" through which the eye could see various environments. The method supported planning, the drawing became a sketch for the future building, visualizations were involved in the practice of engineering. In this chapter visual media that represent human body are questioned as means for discovering or "insight" into the truth and discussed as related with the tendency to gain power over the body. There are three moments of visualization that are closely examined in the first subchapter —the principle of mimesis, the technique of perspective and map making. In the second subchapter the tendency to pierce deeper in the body is analyzed from aesthetic and epistemological aspects. With the birth of the clinic anatomy has started to serve medicine and the gaze of the nineteenth-century anatomist became focal. The focus of the third subchapter is this moment, in which knowledge quite directly came in service of intervention. Yet, there was a delay of intervention into the body because dead bodies were observed in the nineteenth century. This fact prevented timely healing of the ill bodies. Technical imaging on the contrary enables observation of a living ill body. Enormous development of imagery technologies throughout the twentieth century therefore crucially strengthened the power over the body. In the fourth subchapter gaining power over body is discussed in reference to biotechnology. The issue of godlike creating, which is discussed in the first subchapter as regards the principle of mimesis, is here re-examined in the case of synthetic biology. Two terms are central to the discussion: *perspicere* and

© The Author(s) 2017
P. Tratnik, *Conquest of Body*, SpringerBriefs in Philosophy,
DOI 10.1007/978-3-319-57324-3_1

proicere. To be able to comprehend or gain insight (*perspicere*), we *project*. We project what is then to be acknowledged. What then does the quest for the truth mean in this regard? What is this conquest all about? These questions become seemingly even more actual with the intervention of biotechnology because now we manipulate matter according to our own program, we are engineering and designing. In other words, we are creating a world in which we want to live. At the same time, we strive to explore the universe in order to be able to migrate to other planets. Today, the body has become subjected to the technology of power—i.e. biopower—to such a degree that the power over life itself has been exercised not only over the population in general, but also on the fleshy, micro, nano, chemical and visual scale of a singular body. Biotechnology does indeed give humankind a power over life that was not attainable before, though the intervention of biotechnology belongs to a quest that did not first begin in the biotech century. Moreover, it is a part of a quest that has not yet been completed. But this power is not omnipotent. We are able to intervene in certain ways, but this does not mean that we can intervene in every possible way, nor are we able to create a human being from scratch. Is there a grand narrative here about the attainment of a totalitarian power over life that one can detect in the aspiration to this ultimate goal to "create" life from scratch and which is in turn detectable in the "development" of knowledge about life and the technology designed to manipulate it?

The principles *perspicere* and *proicere* are particularly interesting because they enable an approach to epistemology of media. In the case of body imagery, one can recognize an alternation between the two; however, as we are about to see, it is also highly problematic to speak of a regime of *perspicere* without considering the involvement of *proicere* in it. This will guide us through an analysis of a penetrating eye and the organization of space (mapping the body), and it will finally get us to recognize that a regime of *proicere* leads away from scopic regimes and ocularo-centricity and concentrates instead on the outcomes, thus favoring mediation and intervention. For this reason, the corresponding approach is engineering. And it is engineering that has begun to undermine science as a discipline established during the modern era. But first, let us consider what these two concepts actually mean.

We can summarize Vilém Flusser's definition of projection as follows: to use imagination to produce images that connote rather than denote, to project "uncoded magic".[1] *Projection* stands in opposition to *transparency*; the Lat. *proicere*, from Lat. *proicio* (Lat. *pro*—from, for, instead; Lat. *iacio*—to throw) is the opposite of Lat. *perspicere*, from Lat. *perspicio* (to see through something and also to perceive, to distinguish clearly). Among the two scopic regimes in modernity, *perspicere* played an admittedly important role. It is the regime of transparency or visibility that has supported the logic of a gaze penetrating through surfaces. For the beginners of modern sciences, it was an important principle for gaining knowledge. At present, it

---

[1]Vilém Flusser, *Towards a Philosophy of Photography* (London: Reaktion Books, 2000), p. 16.

still has this importance. However, theory may itself "be understood not as a con-templation of form but as a shaping of it."[2] This thought of Flusser's instantly turns the issue upside down—it says that theory is not knowledge about the world gained from the world, not *perspicere*, an insight into the truth, but rather a projection. Thus, let us begin this discussion with the issues of reflection and projection.

1.1   Project Transparency (Constructing the Window)

In his notable study on scopic regimes, Martin Jay acknowledges a regime he calls Cartesian perspectivalism as the dominant scopic regime of modernity.[3] The reigning visual model of modernity is "that which we can identify with Renaissance notions of perspective in the visual arts and Cartesian ideas of subjective rationality in philosophy."[4] One might easily find correlations between the concept of Cartesian perspectivalism and the regime of *perspicere*, but it would be too simplified to say that they are one and the same. As we will demonstrate, the concept of Cartesian perspectivalism is itself not consistent. The idea about the transparency of media, which is part of the essence of Cartesian perspectivalism, was criticized by the visual regime that historically replaced it, the baroque; thus it is problematic to claim that this regime was dominant throughout the whole course of modernity. More importantly, as we will demonstrate, even what Jay refers to as Cartesian perspec-tivalism was not exactly a regime of transparency, of *perspicere*. There was a great deal of conceptual intervention included that could be recognized as *proicere*.

Jay is well aware of the fact that the visual has enjoyed a privileged status throughout the modern era, at least up to the twentieth century, when visuality came under increasing critical attack.[5] As Jay notes, the hegemony of vision is testified as well by Michel Foucault's attention to surveillance, Guy Debord's bemoaning of the society of the spectacle, and Richard Rorty's metaphor of "the mirror of nature" for prelinguistic philosophy.

For his part, Rorty reflects upon epistemology and argues that from antiquity to the baroque one dealt directly with things that exist, thus philosophers were

---

[2]Vilém Flusser, "On Discovery," in: *Artforum*, New York, Vol. 27, No. 10 (Summer 1988), p. 17.

[3]In this study Jay emphasizes that we should recognize the plurality of scopic regimes and detects two others, one that refers to Dutch seventeenth-century art and could be called the art of describing (after Svetlana Alpers), and another which could be best identified with the baroque for its rather passionate approach and its dazzling and disorienting ecstatic surplus of images. In defining the baroque as the second moment of unease in the dominant model, Jay is relying upon Christine Buci-Glucksmann, whose work speaks to the baroque's multiplicities of visual spaces, which are resistant to being reduced to any coherent essence. Martin Jay, "Scopic Regimes of Modernity," in Hal Foster (ed.) *Vision and Visuality* (New York: The New Press, 1988), pp. 2–27.

[4]Ibid., p. 4.

[5]He himself wrote a study about the denigration of vision in twentieth-century French thought. See: Martin Jay, *Downcast Eyes. The Denigration of Vision in Twentieth-Century French Thought* (Los Angeles, Berkeley: University of California Press, 1993).

performing reflections upon the peculiarity of different kinds of entities (from ordinary things to ideas and God).[6] It is plausible that Rorty is paraphrasing Foucault in defining the Renaissance and the classical episteme.[7] As acknowledged by Foucault, it was resemblance that played a crucial role in the knowledge of the Western culture during the Renaissance episteme. Accordingly, that which is depicted is in a direct relationship to the concrete world through resemblance, which is, as Foucault explained, founded in the forms of *convenientia, aemulation,* analogy, and sympathy.[8] The principle of resemblance in this sense holds no awareness of the process of mediation, of the construction of reality with and in media, be it visual media, literature or philosophical reflection. There is no necessary awareness of mediation, just as there is no self-awareness of the comprehending subject. Therefore, the guiding form of epistemological operations in the Renaissance episteme was that of repetition: painting and philosophy reflected nature, a drawing imitated space. But was it really? Could one claim that Renaissance visualizations of the visual world were nothing but copies? What can a copy mean if it is not intended as a re-production, which they obviously were not, since the world was not re-produced, produced again and again in media, but depicted? What are these copies copying then?

Let us discuss three founding Renaissance pillars of visualization that regard the relationship between the world and media: (1) the principle of mimesis, (2) the technique of perspective, and (3) maps, the visualizations of space that were of great importance for Renaissance explorers. As we are about to see, these principles are inter-related, and what is more important, they are very much alive in present-day technologies. Here lie the bases for robotics, medical body imagery and body-related depictions, as in genetic blueprinting, as well as the driving motive for biotechnology, including synthetic biology.

1.1.1 The Principle of Mimesis—Resemblance or Godlike Creation

It is often generally acknowledged that the Italian Renaissance man found the aim of art in mere reproduction—in the sense that the best are those pictures that

---

[6]Rorty summarizes the periods of the history of philosophy and defines the revolution then taking place with linguistic philosophy during the twentieth century: "The picture of ancient and medieval philosophy as concerned with *things*, the philosophy of the seventeenth through the nineteenth centuries with *ideas*, and the enlightened contemporary philosophical scene with *words* has considerable plausibility." Richard Rorty, *Philosophy and the Mirror of Nature* (Princeton: Princeton University Press, 1979), p. 263.

[7]The first English translation of *Les mots et les choses* (Paris: Editions Gallimard, 1966) appeared in 1970, while Rorty's *Philosophy and the Mirror of Nature* appeared in 1979.

[8]Michel Foucault, *The Order of Things* (London, New York: Routledge, 2005), pp. 19–50.

correspond to what they are trying to resemble.[9] However, this notion is not to be found in the Renaissance alone. Resemblance was an important principle in Greek antiquity as well.[10] But what is resemblance exactly? Is it a tracing or a repro-duction (a repeated production)? To trace means there is an original in the concrete world that is optically resembled in the tracing. If there is no physical original, but there is an image (let's say an image of God), then what does such an image imitate? This is not a novel question; it actually takes us back to iconoclasm in the Old Testament's prohibition of visual images. But the question could be re-posed in our own time, the era of biotechnology. Early Christians confronted the following question: because people could mistake the image for what it represents, is it possible and permissible to depict God and produce a material image of Him? When conceptualizing his Republic, Plato expelled image-makers from it because they were imitating imitations (the material world) of the world of Ideas. Philosophers, on the contrary, dealt with Ideas directly; they were the closest to the truth, and therefore they were the ones tasked with managing the Republic. This is well known, but Plato's followers eventually came to different conclusions. Plotinus, a Neo-Platonist from the third century, came up with a different answer to the question as to why people would make images of gods: they do so in order to make gods present in the world and to be able to thereby get closer to them. It was John of Damascus who likewise delivered another great defense of images at the beginning of the eighth century: a holy image shows what is invisible; thus it mediates between the holy and the human world, and it helps people to see God when in fact the appearance of God Himself would make them blind. However, such an understanding was something of an exception. After all, the most intense Christian iconoclasm appeared in the eighth century, but was present also before and after, even as late as the rise of Protestantism. Even the critique of visual culture that took place during the twentieth century, in which we find explicit linkages of

---

[9]This stance has been taken by Arthur C. Danto, and it has helped him to conceptualize his theory of the end of art, within which he generalizes pre-modernist art into a progressive model of the representational line of art (only expressionist art is excluded from this model), which started with Vasari and ended with modernist art. As he claims, this art strove to produce equivalences to perception experiences. See: Arthur C. Danto, *The Philosophical Disenfranchisement of Art* (New York: Columbia University Press, 1986).

[10]The anecdote about the contest between Zeuxis and Parrhasius testifies to the efforts of the painters of antiquity to produce visualizations that could be best described as optical deceptions. It was Zeuxis who depicted the grapes so well that the birds tried to eat them, but it was Parrhasius who won, since he deceived even his rival who was duped into demanding, "Come on, open that curtain and show me what you've painted," when in fact it was the curtain itself that Parrhasius had depicted. One could easily use this anecdote to deconstruct Arthur C. Danto's thesis about art's striving towards increasingly better optical equivalences to perceptual experiences. Danto's narration begins with Vasari, ends with photography and has its postscript in holography. Obviously there was something other than the increasingly more fidelitous production of optical equivalences to perceptual experiences that was of interest to art, even when it had a strong investment in producing so-called copies of the concrete world.

image-making to Platonism,[11] could be understood as yet another chapter of iconoclasm. In the third century an image-maker had a low social status. Tertullian of Carthage considered idolatry an offense of the human species; in his view, it was the devil that produced the images. The image-maker competes with God and thus tries to appropriate His role. Therefore, he is God's opponent. Today we meet quite similar arguments regarding genetics and synthetic biology, in which the question of whether or not we are playing gods reappears.

Greek cultural tradition built upon visuality. It was to this tradition that the Renaissance turned. This, as we will see, has important correlations with the development of technology and science. Renaissance visualizations root in mimesis, which is not exactly the same thing as resemblance. Mimesis is not to be comprehended solely in a technical sense. For the Renaissance artist, the guiding precept here was the recourse to nature. For this reason Leonardo da Vinci was convinced that painting is higher than poetry since it is closer to nature. Nature is a complex of laws, it encloses an essence that one strives to discover and comprehend. The more the Renaissance naturalist strove to resemble nature, the more his ideal was imbued with that of essential form and harmonizing composition. Giorgio Vasari and other Renaissance minds honored this ideal of harmony. Harmony is established with the human selection of natural elements and their principled arrangement into a whole. Planning was therefore of huge importance to Renaissance art. For instance, it is on the basis of a plan that one constructs a building. Planning in this sense means the abstraction and arrangement of things we find in nature. We discuss this issue more in detail in the next section of discussion.

Resembling is not merely a tracing; there is a great deal of rationalization included as well. The preoccupation with harmony (a hallmark of the Renaissance) surely demonstrates another tendency, one that is not directly tied to resemblance. It consists, instead, in a conscious selection and coordination of the parts of the work of art, thus an engagement with forms that will finally compose an effective whole. Additionally, it is important that the pleasant relations remain embedded within the work of art. Therefore, the tendency is not merely to resemble the appearance of nature, but also to compose and arrange a formal whole or to harmonize the arrangement of parts for the purpose of perfection and with the intention that the work of art should please (the spectator). One can trace all of these functions of the work of art back to Aristotle, who requires not just mimesis, but also symmetry and order, which impart beauty and harmony.

---

[11]One of the strongest critics of photography, Susan Sontag, conceptualized her critique of photography in conspicuously Platonic terms and even titled the lead essay of her book *On Photography* "In Plato's Cave". The quarrel between iconolaters and iconoclasts appeared specifically within the Christian context, while the Hebrew cultural tradition originates from speech, hearing and listening, which exceed vision and visuality—therefore, the prohibition of image-making also fits into this framework. The same thing holds for Islam. In the final instance, these differences make differences in cultural products and have great importance for the blossoming of visual arts and culture within the Christian world, while there is not much interest for visual culture within the Hebrew context.

Aristotle, as well as some Neo-Platonists such as Plotinus, provided a complex understanding of imitation, which has less to do with tracing an external appearance and more with probing the internal essence of Nature or even with adding to it, with creating a surplus.[12] How then are we to comprehend Renaissance "resemblance," which was highly influenced by Greek philosophy, especially Aristotle?

Aristotle's notion of mimesis derives from earlier thinkers, particularly from Plato. Katharine Everett Gilbert and Helmut Kuhn recognize the concept of mimesis (imitation)—"not as the duplication of isolated things, but as the active attempt to participate in a superior perfection"—already in the work of the pre-Socratics.[13] For Plato, philosophers aspire to seize the truth, while image-makers who rely upon mimesis are not to be appreciated.[14] Poetry is higher than visual art since poets reflect upon the essence of things. In Plato's view, it all comes down to the relationship to truth, and "painting—and imitation as a whole— are far from the truth when they produce their work; and moreover ... imitation really consorts with an element in us that is far from wisdom, and ... nothing healthy or true can come from their relationship or friendship. ... So, imitation is an inferior thing that consorts with another inferior thing to produce inferior off- spring."[15] Although poets produce their narration through imitation, direct speech within a dramatic performance can achieve the status of a "pure narration without any imitation."[16] Tragedy and comedy are the sort of poetry and storytelling that only employ imitation, while dithyrambs (choral songs to the god Dionysus) only employ narration by the poet himself. After the discussion between Socrates and Adeimantus, tragedy and comedy are admitted into the city, and the kind of mimesis used in this type of poetry is to be allowed because "a single individual cannot imitate many things as well as he can imitate one," and therefore these imitators "must be kept away from all other crafts so as to be the most exact craftsmen of the city's freedom, and practice nothing at all except what contributes to this, then they must neither do nor imitate anything else. But if they imitate anything, they must imitate right from childhood what is appropriate for them."[17] This is a question of style, in this case related to personality, which brings us in turn to nature. If for Plato some arts do not have to use mimesis, then for Aristotle all media rely upon the principle of imitation: a man's action and character are imi- tated, which means that even dancing is the imitation of character and emotion.

---

[12]This moment will later become essential for Hegel's theory of art: art and the artificial have a higher value than bare nature because art encompasses the spiritual and the Idea, which could not be found in bare nature.

[13]Katharine Everett Gilbert and Helmut Kuhn, *A History of Esthetics* (Bloomington: Indiana University Press, 1954), p. 9.

[14]The material world materializes ideas while mimetic images imitate the material representations of these ideas, thus making them second-order imitations.

[15]Plato, *Republic*, trans. C. D. C. Reeve (Indianapolis, Cambridge: Hackett Publishing Company, Inc., 2004), Book 10, 603a10–603b1b, p. 307.

[16]Ibid., Book 3, 384b, p. 75.

[17]Ibid., 396b9–395c4, pp. 76–77.

Stephen Halliwell argues that the concept of mimesis in Plato is not to be taken in an entirely negative way, since even the philosopher is engaged in mimesis ("everything is mimesis and image-making"), and Plato himself wishes that higher kinds of mimesis, those that are closer to truth, be seriously considered.[18] Plato is fond of the notion of mimesis as mere copying of appearances, thus the mimetic artist *par excellence* is the image-maker, the painter. Yet Plato also applies the model of mimesis to the philosophical enterprise itself, which ultimately brings us to a rather complex notion of mimesis.

What later came under the name "arts", i.e. painting, poetry and music, were all to be considered sorts of crafts in antiquity, *technē*. Craft begins with handiness coupled with the impulse to imitate.[19] *Technē* here means the artfulness of mind to trick nature and turn it to man's advantage. This is the semantic origin of the terms technique and technology. For Aristotle, the process of imitation is natural to mankind, and he who is most imitative of the manners and customs of nature learns a great deal through this very imitation.[20] In the end, Aristotle believes that "art" completes what nature has begun. It goes beyond the model only after long schooling according to the model of Nature, in which Nature is energetically working toward a goal.[21] The goal of "art" is to assure pleasure (through the imitation of means, object and manner) as "men find pleasure in viewing representations because it turns out that they learn and infer what each thing is."[22] Only after a long-drawn-out development does imitation become worthy of being called *technē*, that is to say a creation that is in accordance with a true idea. Such a model of art cannot be considered as mere naturalism. A true painter produces an image that is more beautiful than the appearance of what it depicts. Its idea is embodied in beauty. Therefore, the beautiful is above all the spiritual and not so much the appearance. Mimesis accordingly assures not only pleasure (and we want to stress this), but also cognition. To achieve pleasure already means that cognition has taken place as well. One gets to know not only the external appearance of that which is depicted but, most of all, its essence, the idea that is immanent to it and which serves it as model, as ideal. In such a manner, an experience of the art product actually constitutes the "insighting" of the truth.

For Aristotle, mimesis is also central to medicine, which is to be considered a *technē* as well. *Technē* works like nature in the sense that they both subordinate their products teleologically, for the sake of ends; *technē* even completes nature by

---

[18]Halliwell also emphasizes Aristotle's origins in Plato. See: Stephen Halliwell, *Aristotle's Poetics* (London: Duckworth, 1986), p. 119.

[19]Katharine Everett Gilbert and Helmut Kuhn, *A History of Esthetics*, p 62.

[20]*Aristotle's Poetics*, trans. Leon Golden (Tallahassee: Florida State University Press, 1981), chapter IV, p. 7.

[21]Gilbert and Kuhn, p. 62.

[22]*Aristotle's Poetics*, chapter IV, p. 7.

bringing about more than nature was able to accomplish on its own.[23] In this account, medicine and nature are alike, since they produce the same ends in the same way: by subordinating each thing they do to the ends at which they aim. From this Paul Woodruff concludes: "If medicine is to intervene and then let nature carry on the natural process of maintaining health, it must arrange for nature to take some medical artifice as if it were natural; that is, it must produce an artificial effect that merges smoothly in the health-promoting course of nature."[24]

Let us make a short excursus to the phenomenon of robots imitating humankind that is a significant example of the convergence of mimesis and *technē*, one that takes us to the present day. This particular case demonstrates the intertwinement of resemblance with engineering, two poles that perhaps seemed opposed at the beginning of this discussion. For the reasons mentioned so far, and for others that we will soon discuss, it is not surprising that early modern automatons in Europe were produced in the sixteenth century.[25] Automatons reveal that the body is understood in terms of machinery, the principles of which (the mechanics) are to be studied as the craftsman ("artist") has recourse to nature and uses mimesis for his *technē*, with which he might be able to create a body on his own. By the eighteenth century, the interest in robots simulating humankind increased. They became particularly popular in the 1980s with the enthronement of the culture surrounding the

---

[23] Aristotle, *The Physics*, trans. Philip H. Wicksteed and Francis M. Cornford (Cambridge, MA and London: Harvard University Press, William Heinemann Ltd., 1963), Book II, Chapter VIII, 199a, p. 173.

[24] Paul Woodruff, "Aristotle on *Mimēsis*," in: Amélie Oksenberg Rorty (ed.), *Artistotle's Poetics* (Princeton: Princeton University Press, 1992), p. 78.

[25] In the Renaissance the interest in automata actually increased. Complex mechanical devices were known in ancient Greece, and in the 8th century Muslim inventors and engineers produced recipes for artificial snakes, scorpions, and humans (Jābir ibn Hayyān, *Book of Stones*). In his *Book of Ceremonies* (*De Ceremoniis*), Constantine VII Porphyrogenitus (Constantine, 913–959) mentions three automata related to the "throne of Solomon": trees with singing birds, roaring lions, and moving beasts. The western ambassador and chronicler Liudprand of Cremona also alluded to automata of lions and singing birds in the palace in his memoirs of his trip to Constantinople in 949. Several Byzantine chronicles give evidence of automata at the court of the emperor Theophilos (829–842). Furthermore, the Islamic world was fascinated with these fantastic devices. The Abbāsid palaces of the capital of Samarra may have had automata (Muslim accounts mention the amazement of two Byzantine ambassadors to the Abbāsid court in Baghdad in 917 at the sight of a lavish artificial tree with singing birds placed in a pond). In both cultures the contraptions were based on the same principles devised by the engineers of late antiquity, such as the 1st century inventor Heron of Alexandria. In 1206 the Artuqid sultan Nāṣir ad-Dīn Mahmūd ordered a book on automata from his engineer Al-Jazari. In the *Book of Knowledge of Ingenious Mechanical Devices*, the latter sketched and described fanciful devices, such as an elephant clock and a hand-washing device in the form of a servant pouring water from a pitcher that was driven by a complex hydraulic system. See: Mary-Lyon Dolezal and Maria Mavroudi, "Theodore Hyrtakenos' *Description of the Garden of St. Anna* and the Ekphrasis of Gardens," in: Antony Littlewood, Henry Maguire, and Joachim Wolschke-Bulmahn (eds.), *Byzantine Garden Culture* (Washington, D.C.: Dumbarton Oaks, 2002), pp. 128. Around 1495 Leonardo da Vinci designed a humanoid automaton (a mechanical knight) that could independently maneuver its arms, stand, sit and raise its visor. The robotic system was operated by a series of pulleys and cables.

personal computer,[26] when even an ordinary computer was understood as a sort of android able to imitate human brain activities to a certain extent, though not able to move autonomously, hold or move things, listen, watch or feel. The hope to create such a device or at least a part of it has remained an inspiration for numerous researchers in computers and for other technical scientists.[27] Android science now looks closely at the findings of the cognitive sciences, particularly those concerning the interaction between humans and robots. The researchers of robotics today aim at adapting the mechanisms underlying successful inter-human interaction in order to create robots with which people could easily communicate. Yet questions regarding androids remain similar. In 2006 Hiroshi Ishiguru (University of Osaka) developed the first geminoid prototype HI-1. *Geminoid* etymologically derives from Lat. *geminus*, meaning twin, and Lat. *oides*, meaning similarity. Accordingly, the robot is modeled on its creator. The visual resemblance to Ishiguru's appearance is quite good (though not as good as some of the photorealistic sculptures, for instance those made by Ron Mueck); the robot is not able to move in space, it makes clumsy gestures and speaks several languages, which it is even able to use for independent communication with people. The teleoperational system of the geminoid generates autonomous movement of the robot, micro-motions during the process of speech and listening (which differ in both cases), such as take shape spontaneously in human beings. The collaborators in the project propose "to use androids that behave similarly to humans for studying what it essentially means to 'be human', i.e. the mystery of human nature. Androids and geminoids are artificial humans that allow us to investigate human nature by means of psychological and cognitive tests, which we conduct during interaction with people."[28] The Cartesian deception of the senses is actually not to be avoided, but accounted for in a positive sense: "If we could build an android that is very similar to a human, how can we distinguish a real human from an android? The answer is not trivial. While interacting with androids, we cannot see their internal mechanisms and thus we may simply believe that they are human."[29]

Androids are therefore a discernible example of the Aristotelian type of mimesis as it is to be found in contemporary culture. The founding principle of bionics, another branch of robotics that applies biological systems to technology, is again mimesis. As a knowledge-technology that solves technical problems through the study of the functions of living beings, bionics is at present in full bloom in medicine, particularly in the development of prosthetics. Here it is occupied with the question of how to develop the ultimately functional prosthetic limb while paying crucial attention to biological models. The next generation of bionic

---

[26]They were also widely represented in popular culture—see for example the movie *Blade Runner* from 1982.

[27]Peter Laurie, *The Joy of Computers* (London: Hutchinson, 1983).

[28]ATR Intelligent Robotics and Communication Laboratories, "Geminoid HI-1," in: Gerfried Stocker and Christine Schöpf (eds.), *Human Nature. Ars Electronica 2009* (Ostfildern: Hathe Cantz, 2009), p. 221.

[29]Ibid.

prostheses will replace lost limbs not only in the functional sense, but also sensorily. They will enable smooth cyborgian extensions and upgrade our biological bodies with the implementation of mechanics. We can expect to get bionic skin,[30] which will have the ability to sense temperature and touch (human nerves will be connected with carbon nano-tubes arranged along artificial skin comprised of flexible polymers—the active ends of the living nerves will enable sensual perception; the bionic skin will also be equipped with temperature and pressure sensors, and will have implemented artificial hair). Robotics is full of biomimetics, biologically inspired and imitative technology. The very form of the robot is developed with recourse to the body by mimicking its mechanical functions, such as those to be found in muscle, body movements and balance. It has proven to be a particularly difficult objective to develop a robot with the balancing aptitudes to be found among humans, especially for more demanding actions, such as running, playing football and rising to one's feet. At present researchers are aiming to equip robots with a "digital memory" consisting of a digital database collected from the human mind (video recordings from the perspective of the body, taken in the best case during the period of a lifetime) and to equip the digital-mechanical systems of the robot with "wet-brains." These could be biological networks made up of nerves; such an "artificial nervous system" has proven to have the ability to learn (i.e. remember) and act in accordance with these lessons (memories). Or other biological systems could be used that hold some features or qualifications which are not (yet) attainable by mere computer systems. A single-cell organism of a slime mold seems a promising artificial intelligent system as it has proven to provide intelligent, simple and effective (communication) solutions when tested in complex environments, such as labyrinths. Some researchers (like Jürgen Schidhuber) aim to construct the ultimate intelligent organism, a scientist that will be smarter than its (or should we say his?) inventor. The question worth special attention here is the extent to which it is legitimate to refer to androids as "artificial humans." This is actually a question of the technique of mimesis, insofar as it is a question about how far we are able to go with imitating humans and what the status of these imitations is. What are the grounds for determining the status of these human imitations, and what politics are to be applied to them? Antonio Damasio, a neuroscientist, has acknowledged the importance of emotions over the course of an individual's life, especially regarding one's long-lasting relations and inclusion in a social world.[31] We might wonder how successful androids could be in this regard. The other question is a biopolitical one. In a world that is already overpopulated with humankind, do we really need to produce another species, a new sort of "humankind"? The question about creating a robot species patterned after a human model links up with the work of God, who created the human species after Himself.

---

[30]See FILMskin, a common project of the Federal Laboratory at Oak Ridge and NASA that is developing bionic skin for patients with burns.

[31]Antonio Damasio, *Descartes' Error. Emotion, Reason, and the Human Brain* (New York: G.P. Putnam, 1994).

Or perhaps we had better put it this way: man is in the midst of creating a robot, just as he has created God, after his own model, albeit with some improvements.

Leonardo da Vinci claims art must have recourse to nature. Man has no chances to win in competing with nature (in the sense of bettering it), but needs to consult nature about everything he undertakes: "Whoever flatters Himself that he can retain in his memory all the effects of Nature, is deceived, for our memory is not so capacious: therefore consult Nature for everything."[32] Therefore, the best painting is the one that best imitates: "That painting is the most commendable which has the greatest conformity to what is meant to be imitated. This kind of comparison will often put to shame a certain description of painters, who pretend they can mend the works of Nature."[33] However, this naturalistic stance, which we find among Renaissance painters, is not to be taken as one of mere optical duplication. For da Vinci, depicting a human body is much more than just producing a resemblance of the visual appearance of its surface. The method he defends instead is: "Study the science first, and then follow the practice which results from that science."[34] In practice this means that in order to paint a body, a painter has to know its anatomy, composition, and parts, like its bones, joints, skeleton, muscles, etc. Furthermore, a painter needs to know the body in action, its interior (the exertion of certain muscles in certain positions), its external appearance (what the body covered with skin finally looks like and why) in accordance with a given position and action and with a particular body considering its age, whether it be that of a child, a fat man, etc.[35] For Leonardo, to paint is to first conduct a study of how something works and why. Leonardo's observations of the body, which ground the mimetic principle the painter is to use in painting, are therefore scientific: "The flesh which covers the bones near and at the joints, swells or diminishes in thickness according to their bending or extension; that is, it increases at the inside of the angle formed by the bending, and grows narrow and lengthened on the outward side of the exterior angle. The middle between the convex and concave angle participates of this increase or diminution, but in a greater or less degree as the parts are nearer to, or farther from, the angles of the bending joints."[36] It is true that the visual outcome displays the visible world, but first of all it shows the invisible, that which is hidden to the naked eye of the ordinary observer of the scene or person depicted. The artist *knows* more than he can see, he grasps the inner essence, the truth. In the visualizations produced by the painter, the real nature of things is revealed.

---

[32]Leonardo da Vinci, *Treatise on Painting*, trans. John Francis Rigaud (London: George Bell & Sons, 1877), article 365, p. 156.

[33]Ibid., article 351, p. 150.

[34]Ibid., article 27, p. 10.

[35]It is worth noting that for da Vinci there is no such distinction between the interior and exterior of a body as we have drawn here. We have done so in order to emphasize his interest in the whole nature (composition, functionality, etc.) of the body and not only in its visual appearance; that is to say, da Vinci is yet another forceful reminder that mimesis is not at all equivalent to the mere resemblance of only the visual appearance of surfaces.

[36]Leonardo da Vinci, *Treatise on Painting*, article 50, p. 17–18.

The Renaissance's "resemblance" is often generally regarded as *trompe l'oeil*, in which a painting functions as a window through which the observer looks at the "scene out there," external to the painting, i.e. the medium. This notion supports the idea of the medium as seemingly transparent. The eyes pierce through the wall or the canvas and focus on objects in the space on the other side, where it is also significant that these objects are gradually arranged in a space which extends not only towards a horizon of what can be seen, but even further on to where our eyes cannot reach. This is *perspicere*. The medium is denied as a medium; we are looking deeper and farther than we normally would. But it is also a matter of deception: the task of the painter is to deceive the viewer. There is an objective to be achieved, and it is not about unveiling the obstacle of a wall; the painter needs to construct a reality that does not exist at the particular location in which it will appear. The viewer's role in this game is to *recognize*. Actually, there is a very complex process that takes place when we observe these pictures, a process that is based on the unique human ability to recognize and to *imagine*. Not only do we need to imagine Mona Lisa as an existing person, we also need to *see* the objects "behind" her, and in fact we actually see them on the sides of her portrait and we thus *understand* that they are getting smaller and vaguer across the distance of the open space of that scene. However, we likewise *understand* that there must be similar objects *behind* her depicted body, as well as on the left and right side and there, even deeper in the space, too distant from us or hidden behind the rounded globe. There are several artistic tricks used to create these effects, etc. We actually *perceive* much more than what we *see*. There are great parts of the space that are not directly depicted but that are nevertheless *represented* within the painting.

1.1.2 The Technique of Perspective: Putting Order into Relations Between Things

Thus, we cannot completely agree with Foucault (*The Order of Things*) in ascribing the emergence of representation only to the baroque. Even the bare idea of the medium's transparency could not be comprehended as *perspicere per se*. Jay's concept of the Cartesian perspectivalism is additionally supported with the notion of the Cartesian observer, which is comprised of a cold, singular static eye that looks at the scene from afar, as a geometer would gaze at the world, as if he were positioned outside it, not involved in it. However, we could develop two objections to this view. The first objection is related to the critique of *perspicere*, which seems to play the crucial role in this regime. For this regime, the two-dimensional plane was supposedly organized by following the transformational rules spelled out in Alberti's *De Pittura* (1435–36), and later by Dürer and others, so as to render the three-dimensional, rationalized space of perspectival vision. In order to deliver a three-dimensional space in a two-dimensional plane, a particular rationalization or concept of space was used. A few years before the inauguration of the cupola of the

Florence Cathedral *Santa Maria del Fiore* (August 1436), when Filippo Brunelleschi won such widespread public acclaim that Alberti ended up dedicating his book to this genius, Brunelleschi demonstrated for the first time in his two perspective panels how the optical laws of mirror reflection could be applied to painting. Brunelleschi constructed a device with a peephole in the drawing board and a mirror to test his perspectival drawing against the actual view. This technique of mirroring is actually not about opening the opaqueness of the walls or other surfaces to see through them; in other words, it is not *perspicere*, but rather *proicere*. However, Alberti was inspired by Brunelleschi, though in explaining the technique of perspective step-by-step he bases it on the idea of a painting as a window rather than a reflection in a mirror: "I inscribe a quadrangle of right angles, as large as I wish, which is considered to be an open window through which I see what I want to paint."[37] Alberti's instructions go on to demonstrate that the depiction of space on a two-dimensional plane is a matter of anticipating a system of grids. It is the geometrification of space: its measurement, mathematization, and visualization, which makes this a conceptually constructed space. A concept of space precedes its depiction, and then upon the surface this space gets organized mathematically. The technique of perspective is all about getting data points from space and organizing them in certain relations. In short, it is map-making. It also means introducing order into space and space relation. One must gather data from the space, comprehend them, connect them to each other, apply them to the theoretical and technical matrix defining the underlying law for their interrelationships and positions in this space, and finally project this specific rationalization back onto the space as a theory determining our future perception of it. This theorization continues to change our comprehension of the world. Accordingly, Renaissance perspective drawing is to be understood as a predecessor of computer graphics and computer-driven visual simulations. The technique of perspective is all about digitalizing space and programming it, thus transforming every spatial situation into a pool of data, a database that secures the data to be applied to a matrix in which their possible positions are predefined.

It is therefore not at all surprising that linear perspective has been a very important tool for architects. It is a technique that enables the drawing of a sketch for a building that will be erected based on that sketch. Filippo Brunelleschi drew *Santo Spirito*, the church of the Holy Spirit, in Florence in about 1428. The church was built using his sketches (the constructions began in 1436). The erection of the building thus followed his *project*.

The importance of theory for the work of a painter is testified to by Leonardo da Vinci's notes on linear perspective: "Those who become enamoured of the practice of the art, without having previously applied to the diligent study of the scientific part of it, may be compared to mariners, who put to sea in a ship without rudder or compass, and therefore cannot be certain of arriving at the wished-for port. Practice

---

[37]Leon Battista Alberti, *On Painting*, trans. John R. Spencer (New Haven: Yale University Press, 1970), p. 56.

must always be founded on good theory; to this, Perspective is the guide and entrance, without which nothing can be well done."[38]

The other objection to the concept of Cartesian perspectivalism refers to Jay's unification of the Renaissance and classical epistemes. Jay links Cartesian perspectivalism with Cartesian intellectual inspection and refers to Rorty's interpretation of Descartes, particularly because it proves the dominance of vision and visuality in modern Western culture. But according to Foucault, Descartes belongs to the classical episteme,[39] and Rorty likewise detects a similar epistemological break taking place with Descartes. Both Foucault and Rorty analyze how the way of knowing changed with the classical episteme. The classical episteme ceased to operate on the basis of resemblance: "resemblances and signs have dissolved their former alliance; similitudes have become deceptive and verge upon the visionary or madness; things still remain stubbornly within their ironic identity: they are no longer anything but what they are; words wander off on their own, without content, without resemblance to fill their emptiness; they are no longer the marks of things; they lie sleeping between the pages of books and covered in dust."[40] It is representation which becomes the basis for the classical episteme. We find Rorty supporting Foucault in his terms for the representational theory of perception: according to the Thomistic (and possibly Aristotelian) conception, "knowledge is not the possession of accurate *representations* of an object but rather the subject's becoming *identical* with the object. To perceive the difference between this argument and the various Cartesian and contemporary arguments for dualism, we need to see how very different these two epistemologies are. Both lend themselves to the imagery of the Mirror of Nature. But in Aristotle's conception intellect is not a mirror inspected by an inner eye. It is both mirror and eye in one. The retinal image is *itself* the model for the 'intellect which becomes all things,' whereas in the Cartesian model, the intellect *inspects* entities modeled on retinal images. The substantial forms of frogness and starness get right into the Aristotelian intellect, and are there in just the same way they are in the frogs and the stars—*not* in the way in which frogs and stars are reflected in mirrors. In Descartes's conception—the one which became the basis for 'modern' epistemology—it is *representations* which are in the 'mind.'"[41]

However, Rorty's understanding of representation is not the same as Foucault's. Rorty builds his theory on the idea that philosophy mirrors nature. Representation is thus reflection in the mind; whereat "mind" is a separate entity in which "processes" occur, as this notion was established in the seventeenth century, particularly by Descartes.[42] Representation is re-presentation, a doubled appearance. One can thus

---

[38]Leonardo da Vinci, *Treatise on Painting*, article 112, p. 37.

[39]Foucault determines that the classical episteme takes place between the seventeenth and eighteenth century.

[40]Michel Foucault, *The Order of Things*, p. 53.

[41]Richard Rorty, *Philosophy and the Mirror of Nature*, p. 45.

[42]Ibid., p. 4.

question its accuracy. And it is vision that dominates this comprehension: "It is pictures rather than propositions, metaphors rather than statements, which determine most of our philosophical convictions. The picture which holds traditional philosophy captive is that of the mind as a great mirror, containing various representations—some accurate, some not—and capable of being studied by pure, nonempirical methods. Without the notion of the mind as mirror, the notion of knowledge as accuracy of representation would not have suggested itself. Without this latter notion, the strategy common to Descartes and Kant—getting more accurate representations by inspecting, repairing, and polishing the mirror, so to speak—would not have made sense."[43]

The metaphor of the mirror is interesting by itself. While Rorty and Alberti seem to understand the mirror as a surface onto which pictures get imprinted, like a graphic print of the matrix, the mirror nevertheless has an additional feature—that of reflecting the light which reaches it, the ability of thereby casting it onwards, of *projecting*. Brunelleschi seems to know how to make good use of this quality of the mirror. The exploration of the visual was vital to Renaissance artists and researchers. It was during the Renaissance that the body starts being mapped, as was the world. The interest for the mirror arises from this fascination and exploration of the visual, including the processes of perception. Alberti does not yet discuss the complexity of vision, but he does question the location of the images we get from viewing, and it is again the mirror, which might be a part of ourselves as perceiving subjects: the question is whether vision "*resides at the juncture of the inner nerve or whether images are formed on the surface of the eye as on a living mirror.*"[44]

Additionally, as regards the visualization of space, the mirror introduces different perspectives. Thus it is possible to exceed the limitation of a single static eye, gazing from one distant point at the scene on the other side of the frame. With a mirror in this space, we can get a reversed perspective. In the north at about the same time, Jan van Eyck depicted the *The Arnolfini Portrait* (1434), in which the portrayed couple is seen from the front, as well as from the back side. But Foucault sees a crucial difference between the duplicating role which mirrors played in the tradition of Dutch painting ("they repeated the original contents of the picture, only inside an unreal, modified, contracted, concave space. One saw in them the same things as one saw in the first instance in the painting, but decomposed and recomposed according to a different law"), and the role which they played in the baroque, as used by Diego Velasquez in *Las Meninas* (1656): "Here, the mirror is saying nothing that has already been said before."[45]

While for Rorty representation is almost the same thing as resemblance (though he defines the philosophy of the seventeenth and eighteenth century as being concerned with ideas, which brings his theory closer to Foucault), Foucault's theory

---

[43]Ibid., p. 12.

[44]Leon Battista Alberti, *On Painting*, p. 47.

[45]Michel Foucault, *The Order of Things*, p. 8.

of representation is semiological, thus the classical episteme becomes an era of naming, signifying, building signs and constructing order: "The art of language was a way of 'making a sign'—of simultaneously signifying something and arranging signs around that thing."[46] In the classical episteme we are thus dealing with a doctrine of *quid pro quo*. In the visual arts the most significant technique was that of illusion, where the connection between resemblance and illusion was established and fictitious resemblances started to appear everywhere: "Games whose powers of enchantment grow out of the new kinship between resemblance and illusion; the chimeras of similitude loom up on all sides."[47] The baroque is a madness of vision and visibility in which innovative uses of visuality are worshipped. The position of the observer becomes important because the visible depends on the observer's position, as in the form (which is actually the existence) of the cupola depicted by Andrea Pozzo for the *Apotheosis of Saint Ignatius* church in Rome (1685–1694). The baroque is at the same time a critique of vision. Vision and the visual become unreliable. But, again, the perceptual tricks of visualizations that were dependent on the act of observation had already appeared in the Renaissance; consider only Hans Holbein's 1533 painting *The Ambassadors*.

For Foucault, the classical episteme begins with *Las Meninas*, when representation is distinguished from resemblance, that is to say when painting does not represent what can be resembled or depicted, but rather what the whole of the semiological syntagm signifies. In literature the beginning of the classical episteme is marked by Miguel Cervantes' novel *Don Quixote* (1605/15), when the art of literature becomes the art of "making a sign". Don Quixote is a sign, "a long, thin graphism, a letter that has just escaped from the open pages of a book."[48] He is a language, he is composed of words. And he writes himself. He wanders around the world among the resemblances of things. He connects the map he draws with the world. He is an early version of the Pink Panther, the example taken later by Gilles Deleuze and Felix Guattari, who paints the world pink: "The Pink Panther imitates nothing, it reproduces nothing, it paints the world its color, pink on pink."[49] Don Quixote is a rhizome.

Foucault finds a critique of resemblance even in Francis Bacon, "an empirical critique that concerns, not the relations of order and equality between things, but the types of mind and the forms of illusion to which they might be subject."[50] In 1620 Bacon writes: "The human understanding from its own peculiar nature willingly supposes a greater order and regularity in things than it finds, and though there are many things in nature which are unique and full of disparities, it invents parallels

---

[46]Ibid, p. 48.

[47]Ibid., p. 57.

[48]Ibid., p. 51.

[49]Gilles Deleuze and Félix Guattari, *A Thousand Plateaus: Capitalism and Schizophrenia* (London, New York: Continuum, 2005), p. 11.

[50]Michel Foucault, *The Order of Things*, p. 57.

and correspondences and non-existent connections."[51] In our opinion Bacon's stance could actually be read as a critique of mimesis understood as supporting the idea of the order in nature which one discovers and takes for granted (and presents with the technique of mimesis as well) in such a manner that one puts more order into relations between things than might actually exist in nature, i.e. in reality. In such a manner, one presents laws that might not be truly founded. Bacon's critique is oriented against speculative philosophy that presupposes the relations, functionality and order of things, instead of getting information from practical experience, which would testify to what there really is.

There is a connection between the baroque's games of illusion and Descartes' critique of sensual experience as beheld by Foucault. At the beginning of the seventeenth century "thought ceases to move in the element of resemblance. Similitude is no longer the form of knowledge but rather the occasion of error."[52] For Descartes, our judgment about the existence of wax is not grounded in sensual experience, as we do not say that "we judge it to be there from its colour or shape"; instead, we say "that we see the wax itself, if it is there before us." This demonstrates that the wax is not recognized by observation with the eyes, but on the basis of the perception of the mind alone. Sensual experience itself is deceptive: "I look out of the window and see men crossing the square, as I just happen to have done, I normally say that I see the men themselves, just as I say that I see the wax. Yet do I see any more than hats and coats which could conceal automatons? I *judge* that they are men. And so something which I thought I was seeing with my eyes is in fact grasped solely by the faculty of judgment which is in my mind."[53] Thus Descartes does not trust the senses: "I had many experiences which gradually undermined all the faith I had had in the senses. Sometimes towers which had looked round from a distance appeared square from close up; and enormous statues standing on their pediments did not seem large when observed from the ground. In these and countless other such cases, I found that the judgments of the external senses were mistaken."[54] Regardless of the method of proof that is used, Descartes is "always brought back to the fact that it is only what I clearly and distinctly perceive that completely convinces me."[55] It is mind which he trusts: "so long as I perceive something very clearly and distinctly I cannot but believe it to be true."[56]

The quarrel about how one acquires knowledge was as strong between the rationalists as the one between Descartes and the empiricists was in the seventeenth

---

[51]Francis Bacon, *The New Organon* (Cambridge: Cambridge University Press, 2000), p. 42.

[52]Michel Foucault, *The Order of Things*, p. 56.

[53]René Descartes, *Meditations on First Philosophy. With Selections from the Objections and Replies*, trans. and ed. by John Cottingham (Cambridge: Cambridge University Press, 2002), p. 21.

[54]Ibid., p. 53.

[55]Ibid., p. 47.

[56]Ibid., p. 48. Descartes, however, admits the frequency with which his mind is puzzled because he cannot fix his mental vision continually on the same thing so as to keep perceiving it clearly; thus the memory of a previously made judgment may often come back when he is no longer attending to the arguments which led him to make it.

century. Trust in the senses influenced research of the human body and knowledge about it. This brings us to a significant issue: that of truth in the function of the regime of *perspicere*. But before we pay attention to this issue, we would like to briefly discuss the third moment of Renaissance visualization: map-making.

1.1.3 Maps and Exploration—Presaging Exploitation

Measurement is the first act of map-making, intrinsically bound to exploration, which Stephen S. Hall defines as the process of winning data points from nature.[57] Geographical maps of the same area change continually over time, as parts of the world gradually appear on the map and islands become parts of continents (in the seventeenth century, almost a century after exploration had proven otherwise, California was still depicted as an island).[58] In the sixteenth or even as late as the seventeenth century, North America (or the majority of it) remained an empty or borderless figment of the cartographer's imagination. Some sixteenth-century maps included information about its inhabitants and their cultures, signifying either the queerness of the natives or marking the locations of raw material. What was the function of such maps and of maps in general? "If you were a merchant adventurer, you would see new natural resources to exploit, new trade routes to ply, a new geography in which markets and commodities could be paired. If you were the pope, you would see herds of lost souls awaiting conversion, new outposts where missionaries could be dispatched to spread the church's word. If you were the explorer, you would be attracted to those mysterious blank spots just beyond the bounds of knowledge; the map would tell you where your next step should be. All these perspectives converge on a single thematic point, and one that applies no less to modern scientific terrains than to old: shortly after a spatial domain has been mapped, there is a tendency to impose upon it an overlay of values, whether economic or ideological or religious. *Every map presages some form of exploitation.*"[59]

1.2 Deep Body Essence

In 1995 the Visible Human Project (run by the U.S.A. National Library of Medicine) completed a data set of the female body; the male body's digitalization had been completed the previous year. The cadavers of two people had been frozen and cut into thin slices—the male was cut along the axial plane at 1 mm intervals,

---

[57]Stephen S. Hall, *Mapping the Next Millennium*, pp. 8–9.

[58]As many others did, the esteemed British cartographer John Speed depicted California as an island in his 1627 map of America.

[59]Hall, p. 383.

the female at 0.33 mm intervals. The slices were then photographed and digitized: 1871 slices of the male body were collected, which resulted in fifteen gigabytes of data (in 2000 the photos were rescanned at higher resolution, resulting in sixty-five gigabytes of data), and the female scans resulted in forty gigabytes of data. Today, the same work of scanning is going on, but using only parts of the body and gaining higher resolution images.[60] The data acquired by photographing the slices have been supplemented by axial sections of the whole body gained by computed tomography, with an axial section of the head and neck obtained through magnetic resonance imaging and with coronal sections of the rest of the body obtained in the same manner.

We now have a three-dimensional map of the body, of the two prototypes of each sex, actually. In fact, the bodies are a database from which one can restore a few kinds of visualizations of these two bodies. The body has been digitalized. *Corpus* is now subsumed within a pool of space-point information that can rebuild a virtual, i.e. immaterial, three-dimensional corpse. The result is similar to holographic visualization, which is currently extending photography into the third dimension. But here the eye can penetrate the surface and see deeper, it can see *through* the membranes, it can observe the inner structure of the body, its parts, slices of tissues, etc. Using the data set from the Visible Human Project, anybody can use the virtual anatomic construction kit on the web to extract slices, interactively slice through tissue in real-time, construct three-dimensional anatomical scenes combining the slices with three-dimensional models of the internal structures, and other similar things.[61] Yet we are far from the end of this story about the mapping of the body. We are ever improving graphic-technology, the instruments, and knowledge-technology in order to go into even greater detail, which actually means we continue to go even deeper into the body: to explore, map, and conquer it with even greater efficiency.

1.2.1 Anatomical Drawings in Renaissance: "Resemblance" Mediated by Rationalization

Mondino de' Liuzzi (c. 1270–1326), professor at Bologna University, is known as the restorer of anatomy, but during his dissection of human corpses he relied mostly on Claudius Galen's theories, published in a manual titled *Anatomy* from 1316, which were considered dogma until the fifteenth century and were incorrect to quite

---

[60]The scanning, slicing, and photographing took place at the University of Colorado Health's Sciences Center, where additional cutting of anatomical specimens and gathering of data is still in process. An ethical polemic has been taking place regarding the donors of whole bodies to the Visible Human Project. The male cadaver is from Joseph Paul Jernigan, a 38-year-old American who was executed and who agreed to donate his body for scientific research or medical use, but did not know about the project. The donor of the female body remains anonymous.

[61]http://visiblehuman.epfl.ch/, 6-29-2012.

a great degree. Leonardo da Vinci made more than 200 pages of anatomical drawings for this *Treatise on Anatomy* based on his own dissections (1489–1515). But it was the Flemish anatomist Andreas Vesalius who published *De Humani Corporis Fabrica* (*Fabric of the Human Body*) in 1543, after becoming dissatisfied with the early medical texts he had translated and the teaching on the human body that had been passed down from ancient times. He preferred to believe what he saw by himself on the basis of his own investigations. After arriving in Padua, which was then at the forefront of Italian anatomy and medicine, Vesalius became a professor there; upon the completion of his book, he eventually became a physician of Spanish royalty. *De Humani Corporis Fabrica* is considered the first scientific anatomy book. Vesalius invested four intense years of dissection study in order to supply its detailed illustrations of anatomy and the descriptive texts accompanying them. Vesalius corrected Galen's mistakes and visually demonstrated them (for example, he refuted Galen's insistence that the human uterus is horned by drawing the tube-like uterus of a cow and a dog to demonstrate how different they are from that of a woman; also, he deliberately added a dog muscle not found in human anatomy over the collar to demonstrate an error in Galen's anatomy). Vesalius' explanation of anatomy is systematic: he presents the skeleton, musculature, veins (he had difficulties in describing the distribution of the cerebral vessels and depicted the carotid artery incorrectly; the branching of the aortic arch more closely resembles the anatomy of monkeys than that of humans, which demonstrates his dependence at times on Galen's theory), the nervous system, brain (he shows that the brain and not the heart is the center of the nervous system, as had been believed previously), trachea and bronchial tubes, urinary tract (he is aware of the prostate but does not recognize it as part of the male urological system and so does not identify it; he still holds the ancient belief that males and females secrete semen), ovaries (he still believes that female reproductive organs, which he calls testis, are identical to those of a male in form and function), and fetus (his description of the human fetus is not accurate because of his inability to study human specimens; thus his drawings of the placenta more closely resemble the fetal coverings of a dog). He also presents the instruments needed for conducting a dissection. His presentations of musculature and the skeleton are not only vividly depicted, but also include several perspectives with dynamic positionings of the body so that one can get a tridimensional appearance. The figures are depicted in standing, walking or leaning positions, suggesting that the bodies are alive (except the one being hanged). In the background we see the idealized Padua countryside to which some of the figures turn and gaze. The internal organs are depicted within the opening of a classical torso, the cut limbs of which resemble marble statues more than they do the cuts of a fleshy body. But the skin is cut out to make a hole in the body and turned up on the sides. The brain is located in the head, the skull cover is removed and the skin peeled back, hanging over the side of the head, while the face remains preserved. The musculature is presented gradually within several pictures. First the body is almost whole, only the skin has been removed, but it is standing independently and

looking. Later, the muscles and tendons are flayed. Vesalius depicts them as
hanging off the cadaver or sometimes lifted away from the body so that the form
and structure are presented. The flayed tendons and muscles are shown here for the
first time, in a manner emphasizing the importance of flaying: the viewer is enabled
to see both superficial and deep structures at the same time. In this proto-cinematic
spectacle, the body loses its muscles, and in some poses and expressions it almost
seems that it is suffering. Finally, the body is hanged, having already been flayed to
such a degree that the internal organs and the muscles of the torso are removed and
the abdominal diaphragm is hanging separately from the wall next to the body. The
rope from which the cadaver has been suspended, and which has been omitted in
the previous illustrations, is depicted here; it is lowered so that the body is not in a
standing position any more, but is presented as it would be in the act of having been
hung, executed. In the next illustration, the body is devastated; it rests against the
wall of a ruin (functioning as a metaphor for the body, itself a decayed ruin). Here
the cadaver is in a state of a collapse, pointing to a rib cage removed from the body
and positioned beside it on the floor.

Vesalius' depictions show the tendons and muscles, released from one end of
their attachment to the bones and dangling down from the body. They have lost
their function and with it their appearance. On the body they were stretched as
strings and constituted it as a (musical) instrument; dangling down they are contracted and released so that they look like snot. The drawing informs us about
certain features of the tissues. In the body the muscles are attached to the bones;
they are stretched and give it the ability to move. As a whole made up of muscles
and organs, the body is harmonized, it comprises an ideal composition. Yet the
body slowly disappears with its dematerialization in Vesalius' drawings. During the
flaying it loses its parts and starts to decompose. This decomposing is double: it is
the decomposing of the decaying body and it is the decomposing of the body as a
whole, as an ideal. The Renaissance doctrine about mimesis and beauty is attested
to here. With its dissolution, the body becomes less beautiful because it is losing its
harmony and its identity as a body, of which the ideal is that of a living functional
body. This is a theater of the dying body, where dying means decomposition: the
gradual loss of its wholeness and functionality, i.e. its identity as a body. This is
not, however, a particular individual identity, but the very essence of the body as an
idea, or rather the ideal. The use of mimesis is recognizable here: Vesalius is not
only resembling the appearance of the object observed, but is also studying how it
is structured and composed. He is aiming to arrive at the essence of it and to
communicate this essence in faithful depictions. The resemblance involved is not a
mere tracing of the visual appearance of the observed object. It is about conveying
the principles of it as a phenomenon. It is about arriving at the truth. And visualization has the function of explaining what the anatomist has to know. It is a
visualization of cognition. "Resemblance" is highly mediated by rationalization.
And rationalization comes from the process of acquiring knowledge via the practice
of dissection. It is about gaining knowledge from experience.

1.2.2   Mechanical Examination: Anatomy and Microscopy—Penetrating Deeper, Knowing More

Anatomy assures practical experience for surgeons and anatomists, as well as for the research and production of anatomical knowledge. For this reason anatomy is often flagged as the origin of empiricism conceived as the theory of knowledge originating primarily from sensual experience. But as we have just seen, Vesalius is actually following Aristotle's doctrine. Francis Bacon, one of the founders of empiricism who is also regarded as the author of the theory of the scientific method that is grounded in sensual experience (observation) and experimentation, avoids *a priori* propositions. 77 years after Vesalius' *Fabrica*, he wrote: "Men fall in love with particular pieces of knowledge and thoughts: either because they believe themselves to be their authors and inventors; or because they put a great deal of labour into them, and have got very used to them. If such men betake themselves to philosophy and universal speculation, they distort and corrupt them to suit their prior fancies."[62] Bacon distinguished between speculative and operative natural philosophy (*De dignitata et augmentis scientiarum*, 1605), though the term "experimental philosophy" did not appear until 1635. Roughly, speculative natural philosophy develops explanations of natural phenomena without prior recourse to systematic observation and experiment, while experimental natural philosophy "involves the collection and ordering of observations and experimental reports with a view to the development of explanations of natural phenomena based on these observations and experiments."[63] Peter R. Anstey draws a trajectory of experimental natural philosophy anchored in seventeenth-century England. It was favored by almost all natural philosophers (Thomas Hobbes was a notable exception) and was finally officially confirmed by the Royal Society in 1667/8, which stated that it "aimes at the improvement of all usefull Sciences and Arts, not by mere speculations, but by exact and faithfull Observations and Experiments."[64]

Experimental natural philosophy was regarded as a novel approach that emphasized the role of the senses in the acquisition of knowledge. This was further related to the deployment of instruments such as the telescope and microscope. Defenders of the school attacked the epistemic status of the hypothesis, since they regarded it as the province of speculative philosophy: "experimental philosophy emphasized the importance of the senses, constantly appealing to observation and experiment, and it decried the use of mere reason in generating hypotheses."[65]

---

[62]Francis Bacon, *The New Organon*, p. 46.

[63]Peter R. Anstey, "Experimental Versus Speculative Natural Philosophy", in: Peter R. Anstey and John A. Schuster (eds.), *The Science of Nature in the Seventeenth Century. Patterns of Change in Early Modern Natural Philosophy*, Studies in History and Philosophy of Science, Vol. 19 (Dordrecht: Springer, 2005), p. 215.

[64]Quoted by Anstey; ibid., p. 220.

[65]Ibid., p. 238.

Hypotheses were castles in the air, mere speculations, and instead of preceding experimental work, they were to be made subservient to experience; they were to be reduced in experiments and observations. Thus, Newton explained his scientific approach to colors: "For what I shall tell concerning them is not an Hypothesis but most rigid consequence, not conjectured by barely inferring …, but enviced by ye mediation of experiments concluding directly & without any suspicion of doubt."[66] For the same reason, Newton went on to reject Descartes' fictions in the 1680s.

Studying Walther Charleton's investigation of anatomy (*Enquiries Into Human Nature*, 1680) and lectures (from 1683), however, Emily Booth argues that "[e] xperimental proof was not the pinnacle of knowledge in anatomy. A compelling analogy could constitute a stronger proof than could an anatomical demonstration. Analogy held a powerful explanatory role, and observations made in one instance often sufficed as the proof in other (analogous) dimensions of physiology."[67]

Cynthia Klestinec discusses the issue of practical experience in anatomy and claims that it "meant many things in the early modern worlds of craft production, philosophical speculation, experiment, and medical practice."[68] Klestinec pays special attention to different approaches to dissection in the late sixteenth century in Padua and Venice. Girolamo Fabrici of Aquapendente, who took the chair of anatomy and surgery in 1565, inspired his students not because he was dissecting corpses and vivisecting animals, but because he connected anatomy to natural philosophy so that the object of his anatomy lessons were not so much the anatomical structures or connections between them, but rather the coordinated actions of the organic soul. Students wanted to learn not only about the structures of the body and their sequences and connections, but also about the techniques of dissection. They found the process of dissection potentially beautiful or rather beautifiying.[69] Technical skill (*peritia*) was linked to the idea of expertise found in the world of learned surgery. Learned medicine only slowly incorporated learned surgery, although surgeons were ubiquitously present in vernacular health care in the early modern period. At Italian universities students could take a doctorate in surgery, and most medical students had considerable experience with it: "In Italy, surgery was not limited to manual skill; it also included knowledge of theoretical principles underlying medicine, the practices of textual commentary, and specific

---

[66]Quoted in Anstey, ibid., p. 225.

[67]Emily Booth, *"A Subtle and Mysterious Machine." The Medical World of Walter Charleton (1619-1707) Studies in History and Philosophy of Science*, Vol. 18 (Dordrecht, Springer, 2005), p. 171.

[68]Cynthia Klestinec, "Practical Experience in Anatomy", in: Charles T. Wolfe and Ofer Gal (eds.), *The Body as Object and Instrument of Knowledge. Embodied Empiricism in Early Modern Science, Studies in History and Philosophy of Science*, Vol. 25 (Dordrecht: Springer, 2010), p. 36.

[69]Ibid., p. 43. Annually, there was a public anatomy demonstration held in the winter months, but in the late sixteenth century students preferred the tradition of the private anatomy, which was marked by an ebullient reappraisal of *peritia* and its transformation into a virtue of the university-trained practitioner. These private events were conducted on a smaller scale and were rather elite.

operations. In contrast, north of the Alps, where degrees in surgery were not available, the surgeon was consistently aligned with lower trades and the barber."[70]

Robert Hooke, the inventor of the first practical compound microscope (compound meaning it had more than one lens; around 1590 the eyeglass makers Zacharias Janssen and his father Hans discovered that magnification could be enhanced by using two lenses), strongly identified with experimental philosophy. In *Micrographia* (1665) Hooke defended strict (sensuous) examination: "These being the dangers in the process of humane Reason, the remedies of them all can only proceed from the *real,* the *mechanical,* the *experimental* Philosophy, which has this advantage over the Philosophy of *discourse* and *disputation,* that whereas that chiefly aims at the subtilty of its Deductions and Conclusions, without much regard to the first ground-work, which ought to be well laid on the Sense and Memory; so this intends the right ordering of them all, and the making them serviceable to each other."[71]

What particularly astonished the early readers of Hooke's *Micrographia* were the detailed drawings of tiny creatures whose existence and appearance people were neither aware of nor familiar with in the way they were with fleas, flies, leaves, and flowers. Man had thus started to enter into worlds that did not belong to the human scale *per se,* and this has had significant consequences for our comprehension of the world, ourselves included. But it not only extended the worlds we know and changed them, it also gave the human species the power to shape these worlds, manipulate them, and act upon them. The gaining of knowledge with this new technologic power meant exercising control over distant domains. The human species realized there must have been worlds that had not been discovered yet, and it thus set out on a journey to these other scales. Then as now, the purpose of these expeditions has been nothing else than to make a profit. The first step needed to achieve that is and was the conquest of the unknown.

Hooke's understanding of the telescope and microscope is very much of our time, insofar as for him these instruments were bodily extensions, artificial organs. In his work, the human being is already becoming a cyborg: "The next care to be taken, in respect of the Senses, is a supplying of their infirmities with *Instruments,* and, as it were, the adding of *artificial Organs* to the *natural*; this in one of them has been of late years accomplisht with prodigious benefit to all sorts of useful knowledge, by the invention of Optical Glasses. By the means of *Telescopes,* there is nothing so *far distant* but may be represented to our view; and by the help of *Microscopes,* there is nothing so *small* as to escape our inquiry; hence there is a new visible World discovered to the understanding."[72]

The term "microscope" was coined at the turn of the seventeenth century by the German botanist Giovanni Faber from the Greek words *micron* (small) and *skopein*

---

[70]Ibid., pp. 44–45.

[71]Robert Hooke, *Micrographia,* preface, Octavo (CD-Rom edition), 1998 (cop. London: The Warnock Library, 1665), n. p.

[72]Ibid., n. p.

(to look at). In 1663 Hooke observed and in 1665 introduced the term "cell" to describe the microscopic units he found in a section of cork (diameter 49 micrometers). In 1676 Antony van Leeuwenhoek observed protozoa (*Amoeba* and *Paramecium*, diameter 50 micrometers), his lens magnified up to 270 times. In 1850 the microscope was the key to uncovering the theory that all living organisms were composed of cells, as well as to formulating some other theories before and after this. Up to today, the microscope has been one of the most important optical devices for observing the human body.

The microscope enables a specific piercing into the body, which means enlarging a frame and discovering entirely new worlds in it, worlds not seen or known before. Our world accordingly becomes bigger. Actually, it is becoming unpredictably enormous, and this inspires the explorers, since there is continually something else to explore; with better instruments and better technology, new worlds will appear. Science and technology go hand in hand. With regard to the technique of microscopy, we would stress but one issue: it is not the living body that is explored, but rather a piece of the body, i.e. dead or dying tissue. The living body and the living processes taking place within that body cannot truly be observed microscopically. Technically, the body is observed with a microscope by means of the technique of biopsy. The following takes place: a piece is taken from a living system, is isolated and set in such a manner that it can be magnified and studied, and finally is thrown away. The cells might be alive when the sample is taken, but they are actually designed to die, because it is not possible to preserve them and render them back into the organism. Only a selected part of the body is observed, one that can be safely taken (for example, in the early stages of surgery and laboratorial work, it was not safe to perform a biopsy of any organs or tissue), and when observed, it is isolated from its milieu, the functional system in which it once resided. In such a sense, the microscopy of the body is an extension of anatomical observation, it means observing the dead or dying body, but with one important difference. The observation can be performed while the body itself is still alive, thus the findings may help medical professionals to intervene into this particular body, which might currently be subjected to a disease process—something that will soon become the aim of the eye penetrating through the opaque envelopes of the body. Not only are they hiding the inner essence of the body, but they will also soon be recognized as the membranes hiding the locus and focus of a disease. Sight itself undergoes a change as well, it starts to become focused.

With anatomy and microscopy, we find ourselves in the regime of *perspicere*. It could be claimed that the main function of this piercing of the surface, this penetrating deeper and inward into it is to know better, to get to know the essence, to dig for the truth. Machines for visibility like the microscope and telescope helped early researchers to see what was invisible to the naked eye. This power of vision enlightened the world, made it transparent, helped us to map it and subject it, even though it was also opening up new questions at the same time. First, there was such great enthusiasm about the power of vision produced by the telescope that people believed that "no part of the sky escapes our glance,"[73] while today we are aware that we have only begun to explore the universe. In the seventeenth century, when visibility was further improved, the instrument replaced the eye, and vision played a crucial role in observation and science, that is to say vision itself became an enigma.[74] This was also the era in which epistemological questions regarding how one comes to know things gained prominence.

The Renaissance world view was at work revealing the human body, exploring it in a manner similar to how explorers were then exploring the world. The Renaissance man dissected the human body in order to understand the human form. Dissections began to be used to learn and teach anatomy directly from the body instead of doing the same from the book. Investigating the human body by opening it up and analyzing it with a purpose to understand it actually entailed the exploration of the body's geography, and this was performed in a manner similar to or on the basis of the same understanding of how the world was then being explored. The world was explored in order to conquer it; the objective of the conquistadors was to gain as much dominion as possible, to conquer the previously unknown parts of the world. Within this process the explored object was comprehended as the unknown, though not as an empty unknown, but rather as a purposeful unknown, that is to say as something which was waiting to be discovered and which would sooner or later be discovered and get explored. The purposiveness of this unknown, waiting to get discovered, is aimed ultimately at its governance. Therefore, it is not surprising that ethnographers and other humanists were among the first who landed in the new world. They imparted knowledge about the land and its inhabitants, which then aided the conquest of it. The exploration of the human body had been taking place for the same reason: to subject the body and life itself to power. One could raise a

---

[73]Quoted in: Ofer Gal and Raz Chen-Morris, "Empiricism Without the Senses: How the Instrument Replaced the Eye," in: Charles T. Wolfe, Ofer Gal (eds.), *The Body as Object and Instrument of Knowledge*, p. 123.

[74]Ofer Gal and Raz Chen-Morris claim that in the seventeenth century new instruments, such as the telescope and later the microscope, did not offer an extension and improvement of the senses, but rather replaced them altogether. The eye was to become a part of the instrument. Researchers were relying on the authority of their instruments, and therefore the human eye was itself nothing but a flawed instrument. As claimed by Gal and Chen-Morris, for Kepler and Galileo the eye lost its autonomous capacity to observe and assure knowledge—it was instead immersed in nature, a part of the very sorts of things that were to be observed. The telescope, on the contrary, was not bound to the physical world, but was instead mathematical in essence, or so Galileo believed (1623). Magnification is a mathematical relation, and the telescope does the job of magnification, regardless of the collaboration of the eye, irrespective of whether it perceives or not. (Ibid., p. 141) Even for empiricists, the eye and the senses in general had become unreliable and perception deficient, while the instruments became infallible because they enabled insight, they imparted the ability to behold the works of Nature, which man then had the capacity to consider, compare, alter, assist and improve, thus making humankind superior to other species, or so Hooke believed: "It is the great prerogative of Mankind above other Creatures, that we are not only able to behold the works of Nature, or barely to sustein our lives by them, but we have also the power of *considering, comparing, altering, assisting,* and *improving* them to various uses." (Robert Hooke, *Micrographia*, preface, n.p.).

provocative question here: are not artists and philosophers, such as Flusser, excited about biotechnology and its ability to manipulate life, simply in the service of this same conquest, the program of which belongs to biopower? Part of this power over life is medicine. Accordingly, medicine is indeed a knowledge-power over the body, as was recognized by Michel Foucault. Recently, however, this power has gained new dimensions with the body engineering paradigm. In the final instance, inspection of the body will be used for power over human life, over singular bodies and over the mass bodies of the population.

1.2.3   Past and Future Maps

Ambitious human projects aimed at mapping the human body, such as the human genome projects and the visible human project, have not reached their end. The most complex area of the body, the brain, has yet to be mapped. Researchers at Massachusetts Institute of Technology (MIT) are working on copying the "hardware" of the brain, and this is being done in a manner similar to how the body was scanned for the visible human project: the brain is sliced, scanned and finally imaged for a computer simulation. *Connectome*, a comprehensive map of neural connections in the brain, is in the midst of being established. Its researchers aim to determine the structure of our memory, which ought to enable them to digitally bring back information stored in the brain. Today the technology is too expensive and clumsy to be used by individuals, but in the future someone with the proper financial background should be able to get his brain preserved in such a manner. One of the related ideas is that this database could then be inserted into robots, and in such a manner our "souls" would be able to travel from their original biological bodies to mimetic mechanical ones. We argue against such a dualistic and mechanical comprehension of body and life.

We have to admit that the concept of mapping that we still find useful for our reflections on the conquest of the body needs to be adapted to more up-to-date technologies. It remains the case that mapping is a technology for conquest commonly used in several fields, though the map is also a concept of a drawn landscape that lays particular stress upon points of interest. The present paradigm still uses the map for its basic orientation, although the particular kind of map it uses has to be smart, know the user's position, his past trajectory, existing communication networks, transportation routes; it also has to know the user's interests and must be able to predict his future trajectories in these territories. Additionally, the simulation of the user installed in this map can get interconnected with others, and thus we end up with complex simulation systems. In accordance with all of this, the concept of mapping as we have discussed it so far is obsolete in the sense that present and future maps are (and will increasingly be) personalized, interactive, and functionalized in a number of labile ways. Furthermore, they are computerized and programmed so that they generate far more information than a static reviewing map ever could. What we have now are smart maps, computer simulations, intelligence systems that supply additional information to the communicator entering the initial data.

In Europe a huge consortium of scientists is working on the Human Brain Project. Knowledge about the brain is being integrated and transferred to computer models to "simulate the actual working of the brain." As we can read from their web page: "The goal is to bring about a revolution in neuroscience and medicine and to derive new information technologies directly from the architecture of the brain."[75] These are still the present projects aiming to grasp our essence (in the report for the European Commission, one encounters the promise: "we can gain fundamental insights into what it means to be human")[76] and to conquer the human body. The extension of the uses of technology to master the body promised by these projects are tremendous and unquestionably worth discussing. The goals promised by the consortium are to "develop new treatments for brain diseases and build revolutionary new Information and Communications Technologies"[77] that converge with biology. We believe we are indeed entering a new chapter of biopower, one in which the lives of people are not only subject to control, but are interconnected within *connectome*, where everything about us is going to be mastered and manipulable, including what we are, what we know and how we come to know things.

1.3 Focal Gaze

1.3.1 Towards Knowledge-Power—Anatomy in Service of Medicine

With the birth of the clinic, anatomy came into the service of medicine. The exploration of the body conducted through anatomy changed, it became focused; anatomists started to look for the focus of disease in order to study the appearance of its effects and thus the life of its characteristics: what it does to the body and what it is all about. Anatomists paid attention to the diseased rather than the healthy body. If in the fifteenth century anatomists were interested in the prototype of the body (its general map), then in the nineteenth anatomists became concerned with locating and identifying diseases. First maps of an area are always very general and inaccurate, while later cartographers not only notice more details about the terrain, but also produce more specialized maps, such as tourist maps, hiking maps, diving maps, sailing maps, road atlases, ordnance survey maps, etc. The purpose of the map defines its features. The gaze of nineteenth-century anatomist became focal. Even if anatomy did not help with the direct treatment of the disease where it

---

[75]http://www.humanbrainproject.eu/introduction.html. 7-18-2012.

[76]http://www.humanbrainproject.eu/files/HBP_flagship.pdf, 7-18-2012.

[77]Ibid.

appeared, it nevertheless did help with mapping changes in the body caused by it. However, there was a certain paradox in trying to analyze and comprehend the life of a disease in order to be able to speak about how to treat a living body when your knowledge derived from gazing at a dead body. Additionally, one could not observe the development of the disease and its effects, nor the complexity of its expansion, because these effects could usually only be observed in the final phase if the disease was lethal; if it was not, then the internal effects of the disease could not be observed at all. Despite immediate deficiencies that medicine faced in the utilization of anatomy, it has become a crucial tool for acquiring knowledge. This has gone hand in hand with the triumph of experimental natural science.

Foucault detects an epistemological rupture that took place at the turn of the eighteenth and nineteenth century. The episteme changed, the gaze and its comprehension changed, and the clinic was born. Foucault analyzes this moment in which modern medicine fixed its own date of birth, and as claimed by Foucault, it was not a sudden explosive mixture of old practices and logics, but rather "[m]edicine made its appearance as a clinical science in conditions which define, together with its historical possibility, the domain of its experience and the structure of its rationality."[78] The birth of the clinic was enabled by the change of the episteme. Foucault draws a distinction between Descartes' episteme (which belonged to the world of classical clarity), the Enlightenment, and the nineteenth century. Actually, the rupture amounted to an enthronement of empiricism and the experiment as its dominant method. If the classical episteme was distrustful of the senses, in the nineteenth century "[t]he eye becomes the depositary and source of clarity; it has the power to bring a truth to light that it receives only to the extent that it has brought it to light; as it opens, the eye first opens the truth," while "the experiment seems to be identified with the domain of the careful gaze, and of an empirical vigilance receptive only to the evidence of visible contents."[79] For Descartes, however, "to see was to perceive …; but without stripping perception of its sensitive body, it was a matter of rendering it transparent for the exercise of the mind."[80] It was light that preceded every gaze and which was the locus of the ideal, thus also "the unassignable place of origin where things were adequate to their essence—and the form by which things reached it through the geometry of bodies."[81] Not only was light the locus of the essence of things, but also "the act of seeing, having attained perfection, was absorbed back into the unbending, unending figure of light."[82] At the end of the eighteenth century, seeing meant to experience the corporeal, the opacity of it, the solidity and density of things closed in upon themselves. The gaze now "passes over them, around them, and gradually into

---

[78]Michel Foucault, *The Birth of the Clinic. An Archaeology of Medical Perception* (London, New York: Routledge, 2003), p. xv.

[79]Ibid., p. xiii.

[80]Ibid.

[81]Ibid.

[82]Ibid.

them, bringing them nothing more than its own light."[83] In other words, this is the era of the penetrating eye, the eye opening and entering the body as it overcomes the obstacles of the enveloping membranes hiding the truth lying deep inside. This episteme acquires the status of an object; it is the episteme of objectivity, when the gaze is no longer reductive, but is rather that which establishes the individual in his irreducible quality.[84] The truth lies in the dark center of things. The empirical gaze has a sovereign power to illuminate things that once were lying in the darkness: "Rational discourse is based less on the geometry of light than on the insistent, impenetrable density of the object, for prior to all knowledge, the source, the domain, and the boundaries of experience can be found in its dark presence."[85] Thus, passive gazing is linked to absorbing experience and mastering it. In terms of our discussion, this is finally *perspicere par excellence*.

In the middle of the nineteenth century, we find a defense of experimental anatomy by Claude Bernard, who practiced animal vivisections. Bernard writes how man can observe the phenomena that surround him only within very narrow boundaries, in which most phenomena escape his senses, meaning mere observation is not enough: "To extend his knowledge, he has had to increase the power of his organs by means of special appliances; at the same time he has equipped himself with various instruments enabling him to penetrate inside of bodies, to dissociate them and to study their hidden parts."[86] Man does not limit himself to seeing, he also needs to learn, which is enabled by observation. Then he can reason, compare facts, question them, and test one against another: "In the philosophic sense, observation shows, and experiment teaches."[87] Nineteenth-century researchers of the human body dreamed of having such an opportunity as had William Beaumont, today considered the father of gastric physiology, who observed the digestion system of a live person and thus had a living human test tube as it were.[88]

Foucault uses the term "anatomo-clinical medicine" to refer to anatomy that serves the clinic by assuring knowledge about disease. Anatomo-clinical medicine consolidates the medicine of visible invisibility, Foucault acknowledges, referring to Xavier Bichat, anatomist and physiologist from the late eighteenth and early nineteenth century: "Truth, which, by right of nature, is made for the eye, is taken from her, but at once surreptitiously revealed by that which tries to evade it. Knowledge *develops* in accordance with a whole interplay of *envelopes*; the hidden element takes on the form and rhythm of the hidden content, which means that, like

---

[83]Ibid.

[84]Ibid., p. xiv.

[85]Ibid.

[86]Claude Bernard, *An Introduction to the Study of Experimental Medicine*, trans. Henry Copley Greene, A.M. (New York: Henry Schuman, Inc., 1949), p. 5.

[87]Ibid.

[88]Beaumont observed the digestion process in his patient Alexis St. Martin, who was accidentally shot and got a hole (fistula) in his stomach that had not closed. Thus, Beaumont got and used the opportunity to conduct a series of experiments with the digestion of food during the normal conscious states of his patient.

a *veil*, it is *transparent*: the aim of the anatomists 'is attained when the opaque envelopes that cover our parts are no more for their practiced eyes than a transparent veil revealing the whole and the relations between the parts.'"[89]

1.3.2 Detailed Maps of Generalized Bodies

Does the same episteme supporting empiricism founded at the end of the eighteenth and beginning of the nineteenth century, praising "the modesty of its attention, and the care with which it silently lets things surface to the observing gaze without disturbing them with discourse,"[90] still ground today's anatomo-medical discourse? The Visible Human Project has succeeded with the task (described by Bichat) of overcoming the opaque materiality of the body better than this was ever possible to achieve with classical anatomical dissection, not to mention with the disturbance of body decomposition, which has effectively vanished. The tactility of matter has been transformed into a digitally opaque obstacle for the eye that can be easily overcome with the help of a computer program. The envelopes of the body have become immaterial veils that can not only be easily unveiled, but also re-veiled by a few clicks of a mouse. There are of course certain differences in comparison to the anatomical cut. For one thing, the body is not actually present. The whole presence of the body—its smell, its tendency to decompose, the intimacy of the particular individuality of the corpse, the whole of the anatomical experience in fact—is absent. There is a break in presence. Furthermore, the performing of the cut is differentiated from the gazing eye. The dissection and the observation are performed by different subjects. The tactile penetration into the body and the penetrating eye do not belong to the same subject, nor do they even belong to the same act; they are not identical to each other, there is a *différance* at work. The observing and knowing subject is absent in the moment of dissection, as well as in the moment of inscription, which involves the transfer of the information taken from the body into digital data. The "anatomical" cut was made prior to this "anatomical" observation, and it was in truth not an anatomical cut at all—it was an overall cut through the tissues, a completely technical cut, irrespective of the body's features, the structures of the tissues and organs, by traversing them all on the same mathematical plane. Accordingly, the structure of the slice is not comprehended as a structure of various matter compositions, some more solid than others, but read instead mathematically, as points of specific qualities on a two-dimensional plane. Finally, these points of each plane are composed into a column, dense with points that can be further interconnected. Only now these data points can be interrelated in such a manner that an educated eye can recognize the structures and forms of the tissues. Meaning starts to arise from this block of digits; semantics can now be laid

---

[89]Foucault, *The Birth of the Clinic*, p. 166.
[90]Ibid., p. xix.

onto this graphical grid. It is important that the graphic be readable, that there is a code which can be detected and used by the reader. But the process of reading itself is delayed; it can happen at any time in the future, but not in this precise moment of inscription. The Visible Human Project produces a map, which is itself a differential mark, a kind of writing; thus, the same essential predicates hold for it as for the determination of the classical concept of writing.[91] The same predicates hold true for the majority of body imagery, including X-raying, MR, CT, and others.

In this particular case, we cannot ignore the fact that the aim of the Visible Human Project is "the creation of complete, anatomically detailed, three-dimensional representations of the normal male and female human bodies."[92] We are surpassingly dealing with sex *representatives*, with *normal* bodies. This is about the norm, standard, and not about individualized bodies. This kind of comprehension was also found in Renaissance anatomical dissections, which were performed with the aim of learning about body composition in general, i.e. about *normal* body composition. The body depicted by Vesalius was also a representative body. Conspicuously it was not a sick body, nor that of an older man or child; instead it was a well-developed male body. Probably it was built up from several dissected bodies. Vesalius' body testifies to the Renaissance norm, to the harmony of the body, to *beauty*. It was an *ideal* body. The scanned body of Joseph Paul Jernigan is not meant to be his particular body (individuality) but rather its generalization. How are we to come from this particularity, linked to a particular name, to total universality; how is it that individuality can become a prototype? How can one particular body (not even an ideal body) become a common human profile, a general identity? The upgrading of the Visible Human Project through the collection of scans taken from parts of the bodies of various people demonstrates that an embarrassment with signified singularity has led to an effort at correction. The ethical problems with revealing the donor's identity and agreement are thus supposedly resolved. Perhaps there still could be donors who are not necessarily aware that their bodies or body parts are being used for this purpose, but there is an excuse that their identity has been hidden, since these parts of the body are hardly identifiable. From our perspective, the other problem is much deeper for it is nothing less than the problem of *singularity*. The logic used in the project is that we can speak about the species by speaking about a singular representative; supposedly this could be just about anybody, the criteria being only that he is not a child, an old man, a woman (if we want a picture of a male), an ill individual, a physically deformed individual, etc. Actually, there are many criteria one has to fit to be a possible candidate for visualization as a representative of the species. We are actually dealing with aesthetic ideals. So many characteristics have been excluded as possibilities in the construction of this body that what is acceptable and thus what is *normal* have been quite clearly defined. Obviously, the authors of the project recognized the body of Joseph Paul Jernigan as

---

[91]We are of course referring to Derrida here. This issue will be discussed in more detail in the next chapter.

[92]http://www.nlm.nih.gov/research/visible, 6-29-2012.

one that was suitable for this purpose. Still, this is Joseph Paul Jernigan, a particular singularity; it is not just anybody or a "general identity" (how could such a strange idea even exist?), even if so many people would want this to be true. This particular semiological network is linked to the semantics of this particular individual: Joseph Paul Jernigan.

The data published as the results of the Human Genome Project do not represent the exact sequence of any one person's genome, but rather a combined profile of several anonymous donors. The authors of the project seem to avoid the ethical problem of identity, but the problem of the common human profile remains the same, even if there is no concrete specimen attached to the signifier. In this sense, then, this is yet another project revealing the essence of the human as a species, the general essence, the essence of us all. What essence could that then be? It is a paradox to say that this is the very essence, a concrete profile, while at the same time claiming that this grapheme, this mark, does not belong to any specific signified. Is it a signifier at all? Because if it is not, then one can hardly broach the issue of coding. And it is coding that we are working on; we are trying to arrive at the code of life, the codes of illnesses and other genetic determinations, because knowing these codes will give us a possibility of intervening into them. We require this knowledge for conquest so as to be able to master its bodily expressions.

Renaissance drawings of anatomical bodies referred to just such a generalized body identity. If one were to compare anatomical drawings with the open dissected body, it would at once become clear that these drawings do not really support the *perspicere* regime, as the perspective drawings or pictures of the beautiful bodies portrayed are not windows. All of these depictions are maps, planes, visual explanations of the bodies. Conversely, the concrete open body is a mess of tissues, solid and liquid matter. The illustrations originated from perception, but the educated eye is able to read this structure and organize it into a map, into a drawing of it. Everything that would in a concrete world disturb the vision and explanation of the territory is purified in the illustration, just as the area at issue is systematized. These are visual theories about body composition.

With the attachment of anatomy to medicine, the observed body becomes individualized. Individualization appears with the particularity of the disease, and there are only individual illnesses because "the action of the illness rightly unfolds in the form of individuality."[93] The anatomical approach of the doctor is thus an approach to an individualized body. And herein lies the very difference in anatomical approaches. Now the observing eye is looking for particularities. The observing eye is an educated eye, with which the doctor analyzes the disease. This is not naïve, naked gazing. It is a focused observation. And the graphic mediation of the computed technology enables the intensification of particular spots that are important for medical examination in order to produce an accurate diagnosis and treatment. Additionally, the technical quality of these graphic visualizations of various kinds is increasingly improving, thus enabling us to observe in ever greater

---

[93]Foucault, *The Birth of the Clinic*, p. 169.

detail. In combination with the development of technical instruments for magnification, the eye has been successfully penetrating deep into the body; at this moment, in fact, we are able to see down to the nano-scale.

Such non-human scales do not ensure a rich sensorial experience. "Experience" itself changes. It is significantly *reduced*. Not only are there no other senses involved besides vision, even vision itself is very restricted. The scope of observation gets framed and particularized—it is a small area that is observed, while the majority is left out of consideration altogether. Additionally, tissue is not observed in the body and is not same as it exists there because it has been transformed and prepared for the act of observation itself. Furthermore, the magnified detail of the human body does not "look like" the human body. Recognition of the body on the basis of the representations we have learned from our everyday perception of the body fails when faced with other scopic scales. However, with new experiences gained from new ways of seeing the body, body imagery and knowledge about the body itself change as well. Our idea of the body gets upgraded and expands; it is in the process of becoming everything that has yet to be added to the basic idea we once had of it, if we dare to simplify this in such a manner. The idea of the body is thus never completed, it is always in the process of becoming.

Additionally, the body is losing its wholeness and consistency. It is dissolving. Furthermore, with microscopy and the majority of graphic visualizations, space has become flattened and distanced from the observer—the object of observation is taking place in front of the observer in such a way that the object and the subject cannot connect and intertwine with each other. This is not a space in which one can become immersed, but rather it is a display. The observer has been excluded from this world. Still, we have prolonged our eye to such an extent that we can now gaze at other scales. And, as we will see, we have even succeeded in touching the objects in these other worlds; we have developed very pointed touching, i.e. techniques for intervening at these other scales.

1.3.3 Contemporary Medical Imagery—Focal Gazing at (a Representation of) an Ill Living Body

The inventions of technical devices that have enabled us to observe the interiority of the human body have become crucial instruments for medical examination. The development of imagery technologies was enormous throughout the twentieth century. X-raying, invented in 1895 and immediately used in medical diagnostics (a role that it has maintained ever since), still continues to improve. X-ray images show solid structures within the body, such as teeth and bones. But X-raying is an invasive technology in that it subjects an individual to ionizing radiation. Immediately after World War II, new technologies of measurement began to emerge and entered clinical use in the 1970s and 1980s. Physicians developed technologies for visualizing, categorizing, and mapping the interior of the human body with unprecedented detail and resolution. Standard X-rays do not distinguish between soft tissues, such as the heart or liver, and they only produce a two-dimensional view of the anatomical terrain. Thus, researchers began to consider more sophisticated uses of X-rays—a new application of which began to emerge in the 1950s. It was called computed tomography (CT). Allan M. Cormack was experimenting with encircling a specimen and firing pencil-thin beams of gamma rays at it from various points of this circle when he realized that internal variations of density and thus structure could be inferred. This inference was mathematical, requiring a computer to assess a larger quantity of equations in order to reconstruct a three-dimensional cross section of tissue. CT scans reveal a covert landscape of malignancy, particularly in the brain; they also excel at revealing the liver, spleen, pancreas, and structures surrounding the heart. In the early 1970s, researchers discovered that organic tissues (biopsy samples from rats were used first) possess a magnetic resonance signature as distinct as a fingerprint. By shooting a pulse of radio energy at tissue, you get small but discernible radio signals emitted from atomic magnets. Once nuclear magnetic resonance could distinguish between biopsy samples in the pathology lab, the next step was to try it on a living human as a diagnostic tool, which happened quickly. The technology of magnetic resonance (MR) is noninvasive (it does not use ionizing radiation), and it produces cross-sectional images. The body can get "cut" anywhere one wants. The technology displays a contrast between flowing and static things and better contrast resolution between diseased and normal tissue than other existing imagery technologies. Thus, differences can be observed earlier. MR enables good observation of soft tissues as well. This is also a technology that enables the construction of a whole body magnetic-resonance-imaging-map (it was a supporting technology in the Visible Human Project, as was CT), which can be digitally stored and compared with cartographic updates years later. These new technologies of diagnostic imaging do not just include the well-known MR and CT, however. For instance, video thermography measures variations in temperature and can pinpoint areas of elevated temperature in the skull, locating the migraine headache nests that set off waves of pain. A sonar device uses ultrasound vibrations to map the inner topography of blood vessels and arteries, showing the location and extent of blockages (also referred to as an acoustic microscope). With an application of Doppler techniques, cardiologists can measure and map the rush of blood through the heart. By the adaption of stereophotogrammetry, orthopedists have mapped the three-dimensional landscape of cartridge and bone.[94] In the mid-1970s, when ultrasound instant images became available, noninvasive technology using high-energy sound waves to probe biological structures started to be regularly used in obstetrics and gynecology. Ultrasound creates an image by measuring the intensity of sound waves bouncing off an object; then a computer, instantly assigning shades of gray or color to these intensity levels, creates an image: of a fetus or a beating heart, for example. Angiography is a technology that inserts a microscopic camera inside

---

[94]Hall, p. 143.

blood vessels. With a procedure called cardiac catheterization, it is possible to locate critical cardiac landscapes and examine them for evidences of blockages of circulation. With nuclear medicine radioactively labeled markers are injected into the body, and these migrate to specific locations and emit signals that can then be picked up by detectors outside the body. Gamma cameras supply full-body images of the skeleton with the use of a radioactive pharmaceutical that migrates preferentially to bone matter—the spread of cancer from primary sites to the bone can thus be checked. Positron emission tomography (PET) uses short-lived powerful positron-emitting isotopes, which are injected into the body and then can be picked up by sophisticated detectors.

There is an important difference between imagery technologies and anatomical dissection, even in the case of the two bodies in the Visible Human Project. In the case of the imagery technologies just presented that are routinely used in today's medicine, there is such a thing as gazing at a living body, only this view is complexly mediated since the eye is actually gazing at a representation of a living body. But the ability that we now have is to open the body without a cut. Neither anatomical dissection nor surgery is needed. Some of the technologies even supply (almost) real-time images. Not being able to look into the living body and having to wait until the patient had died was a great deficiency for nineteenth-century doctors trying to learn about diseases. Accordingly, they were not allowed to see the functioning body, the living body and the life of the disease. Also, because examining dead bodies in order to treat living ones makes little sense, during the twentieth century technologies have been developed that enable us to observe a functional body, a body in action; some of these technologies even provide real-time images. This surely marks another epistemological rupture. Now the object of observation can be the subject as well.

We are penetrating deeper and deeper into the body. But we are also able to grasp several very different perspectives on the same internal situation, on the same scale. One might claim that all of these body imagery technologies are good examples of the *proicere* regime. However, their distinctions already demonstrate that they are not all the same naïve or naked insight into the body, meaning they are not only *perspicere*. The variety of possible imagery attests to the varieties of gazes we have acquired. The object observed does not have only one appearance, but rather a whole variety of appearances. If one were to compare the dissected body with the images collected using present technologies, it would immediately become evident that the observed object has become enriched. We can no longer speak about the plain body, but must refer instead to a *surplus body*. However, this richness of appearances is not an empty surplus. It has been ensured by a focused gaze. The majority of the images of the same location in the body might be useless for the particular purpose a particular gaze might have.

The specificity of each of these technologies is linked to their purposiveness. This purposiveness is necessary, insofar as a particular problem calls for a particular technology of observation. Linked to its medical application, the aim defines the graphic. The graphics are used again and again for particular cases, always according to the aims of observation. These images communicate particular issues;

they answer particular questions, very conceptual questions in fact. They ensure information, which, for a proper reader, becomes cognition. The proper reader is an educated reader, a reader with prior knowledge about the body and disease as well as of the proper code needed to decipher the image. This process of communication between body and reader through the mediation of the body image is what finally instructs the reader as to what proper actions need to be taken in the future.

The body has become optionally transparent, depending on the choice of technology that is made, which itself depends on the particularity of the medical interest. Its solidity and opacity have dissolved. The wholeness of the body has been broken up even more by nanotechnologies. Such technologies have developed with the challenge to pierce even deeper into the human body. However, they have been developed not only with the aim to observe on the nano-scale, for *perspicere*; nanotechnology also aspires to act on this scale, to manipulate matter, the world, the body, etc. Nano-scale is definitively invisible to the naked eye. One nanometer is about half of the diameter of DNA's double helix. This miniature world became visible when it came into existence for our conceptual world in the 1980s with the Scanning Tunneling Microscope. We have already used some nanotechnology before, but not in a way we could fully understand then (in the 1920s car tires were treated with a material called carbon black to cut down on wear and tear; these carbon particles were nano-sized). There have been some successful cases of medical treatment with nanotech drugs so far (for example, nano-particles that deliver a drug to the locus of a disease, e.g. the location of a tumor). It does not work with a dead body, but rather with a living one. The prospects for this technology prove particularly promising in directed medical treatment. Significant for this knowledge-technology is the fact that from its very beginning is has not been developed as a mere technology for observation, but has instead been understood as an instance of *engineering*.

1.4 Gaining Power Over Body

For Flusser, a screen is a two-dimensional projection that we can never penetrate.[95] The theater represents the world of things through things themselves; film represents the world through the projections of things. From this perspective, neither genetic engineering nor a genome sequence is an instance of *perspicere*. DNA visual displays of a profile are a form of transfer, translation, and projection onto another carrier, into another material. However, the regime of projection here does not mean just the technical transfer of a particular picture, but also its active transfer, which is even more than an intervention; it is "throwing onward," planning. *Proicere* does not follow from a central perspective; it is a breaking point, a rebound, a reflection. *Proicere* is thus a critique of the objective, static world,

---

[95]Vilém Flusser, *Writings* (Minneapolis, London: University of Minnesota Press, 2002), p. 24.

independent of our intervention. Maurice Merleau-Ponty criticized Descartes' conception of space and claimed that space is not, as it was in the *Dioptrics*, "a network of relations between objects such as would be seen by a third party, witnessing my vision, or by a geometer looking over it and reconstructing it from outside. It is, rather, a space reckoned starting from me as the null point or degree zero of spatiality. I do not see it according to its exterior envelope; I live it from the inside; I am immersed in it. After all, the world is around me, not in front of me."[96] Byzantine painting, and later cubism, did not use linear perspective to withdraw the objects from the gazing subject, placing them "on the other side" in a distant world; on the contrary, in reverse perspective the "vanishing" points are placed in front of the painting, thus the object is opening up and expanding from the point of the observer. The world is embracing me, however it is still intact by my actions; I remain just an observer. In *proicere* I am the null point of spatiality. But I am not only passively immersed in the world; producing and distributing technical images or with any act of intervention, I *project* into it.

1.4.1   Genetic Program of Life and Intervention

In the 1990s, the computer paradigm had such a great impact on biology that "[t]he organism has been translated into problems of genetic coding and read-out,"[97] as Donna Haraway once put it. There was great anticipation related to the exploration of the human genome. It was to deliver knowledge about what we are and to solve the "riddle of life," if we may borrow a phrase from those days. Within the Human Genome Project, the sequence of the base pairs that make up human DNA was determined and between 20,000 and 25,000 genes of the human genome were identified and mapped. A rough draft of the genome was finished in 2000, and the project was completed in 2003. The project began in 1990 (it cost three billion dollars and was financed by the U.S.A. Department of Energy and the National Institutes of Health), but this ambitious idea about mapping and sequencing the entire human genome was already being discussed in 1985 at the University of California in Santa Cruz by David Botstein and other notable biologists of the time. Likewise, the method of genome sequencing (i.e. a process of writing down the order of DNA nucleotides in a genome in the order of the base pairs A, C, G and T that make up an organism's DNA)[98] was developed even earlier, in the mid-1970s

---

[96]Maurice Merleau-Ponty, "Eye and Mind", in: Galen A. Johnson (ed.), *The Merleau-Ponty Aesthetics Reader. Philosophy and Painting* (Evanston, Illinois: Northwestern University Press, 1993), p. 138.

[97]Donna Haraway, "The Cyborg Manifesto. Science, Technology and Socialist-Feminism in the Late Twentieth Century," in: David Bell and Barbara M. Kennedy (eds.), *The Cybercultures Reader* (London, New York: Routledge, 2000), p. 303.

[98]Cytosine, guanine, adenine (DNA and RNA), thymine (DNA) and uracil (RNA), abbreviated as C, G, A, T, and U.

(in this regard, notable is the work of Fredrick Sanger, who completed the method of genome sequencing in 1977). Even if the book of the body has been written down, it is still quite far from being readable by us and even farther from being mastered, i.e. completely manipulable. The idea of "reading out the book of life" is problematic from several aspects. It presupposes in the first place, that there exists something as plain, simple and static as is a book of life. Not only do we not yet know the code for this language, it is also very doubtful if it can ever be understood as a language. The latter would mean that what we are (our essence) could be written down in the form of a language, in the form of a chain of signifiers which have a corresponding underlying chain of signifiers. The very representation of the order of the DNA nucleotides in a genome, which can be written and supposingly "read out," supports a notion about the language of a species (in this case the human species), where letters and words are interchangeable. Thus, this is a language of life (of all species), which is comprehended in pretty much the same sense as Ferdinand de Saussure comprehended (classical) language. There is a binary arrangement of signs, with a connection between signifier (the signifying element) and signified (the signified content). Accordingly, there is an idea that the organisms are just big containers for this genetic database. This is a comprehension of life as content, software, genotype, which only materializes itself in hardware realization, i.e. phenotype. In other words, it rests upon an obsolete comprehension allowing the separation of form from content; it is another version of Cartesian dualism between the body and the soul. Haraway herself notices that with genetics the body is recognized "as an artificial intelligence system, and the relation of copy and original is reversed and then exploded." From this perspective: "Genesis is a serious joke, when the body is theorized as a coded text whose secrets yield only to the proper reading conventions, and when the laboratory seems best characterized as a vast assemblage of technological and organic inscription devices."[99]

The idea of "reading out the book of life" thus rests upon a belief in genetic determinism. Even Flusser, who in his final writings on science discussed genetic engineering with a great deal of enthusiasm, agreed that we are socially determined for the most part. Last but not least, the idea of the book of life presupposes that there is one book, *the* book, the holy text in which truth is ultimately encoded. Obviously, we have not yet rid ourselves of the idea of the one and only (detectable) truth. Such a belief is likewise supported by a belief in such a thing as proper reading. This is a romantic hermeneutic principle, according to which the work can be understood in accordance with the author's understanding and, even better, with the critique of this traditional hermeneutics as it was conducted by Hans Georg Gadamer.[100] The idea of one proper reading of the "work of art" was of course much criticized by post-structuralists, especially Roland Barthes, and within cultural and media studies.

---

[99]Donna Haraway, "The Biopolitics of Postmodern Bodies. Determinations of Self in Immune System Discourse," in: Shirley Lindenbaum and Margaret Lock (eds.), *Knowledge, Power, and Practice: the Anthropology of Medicine and Everyday Life* (Berkeley: University of California Press, 1993), p. 367.

[100]Hans Georg Gadamer, *Truth and Method* (London, New York: Continuum, 2006).

The whole critique of such an idea about a proper reading which guides a reader to truth has been present throughout twentieth-century continental philosophy, starting with the work of Friedrich Nietzsche and his stance against positivism: "Against positivism, which halts at phenomena—'There are only *facts*'—I would say: No, facts is precisely what there is not, only interpretations. We cannot establish any fact 'in itself': perhaps it is folly to want to do such a thing."[101] This criticism could be applied to the idea about the "genetic book of life," which rests upon a romantic belief in such a thing as a proper reading or tone that will bring salvation. Within genetics, there is a belief that this field is at work in understanding the genome not only "properly," but also "better" than its author, be it understood as nature or God. In this particular case, the plan is as ambitious as possible—to decode life and comprehend its creation in order to finally be able to gain the mastery of doing the same and better for ourselves. Therefore, it is not surprising that the phrase "playing God" is often attached to the ambitions and already to some of the attainments of (molecular) biology or biotechnology.

Sequencing the human genome is not truly about *perspicere* into the body, even if in this case the idea is to penetrate down to the utmost inner essence of the body, to the essential truth of life. Sequencing the genome means writing down the complete set of genetic or inherited information stored in each of the organism's cells. The visual results themselves are already translations into other languages, letters that we understand or digits which constitute the graphic picture. Actually, all of the body imageries mentioned above are translations of the gathered information into another organization of this information. In this sense, they are all already maps. But this particular map (the map of the genome) is the map of maps. Everything should be written in it. We only need to learn how to read it out. The difference between the maps previously discussed and this one is that those were all visual maps, thus the idea about *perspicere* into the body seemed "natural." In a little while we will deconstruct this notion. But in the case of the genome sequence, the results are visual in a different sense; it is as if one were to discuss the visuality of writing. Why? Because obviously here we are dealing with a "highly encoded"[102] semiological network, be it the determination of the DNA nucleotides in the form of chain of letters (the order of the base pairs A, C, G and T) or in the form of a graphic array of the digits composing a chromatographic output.

---

[101]Friedrich Nietzsche, *The Will to Power*, trans. Walter Kaufmann and R. J. Hollingdale (New York: Vintage Books, 1968), Book 3, 481, p. 267.

[102]If language is a highly conventional form of communication that links signifiers to signifieds, then we can no longer speak of convention here, since this is a discourse that was not established by the human species and thus it was not man who established the code for this language. Yet, the code is very complex, thus we could say that the semiological network is "highly encoded." It is worth mentioning that genetic engineers, including synthetic biologists, still believe they need to figure out the language with which they might program DNA. They need to learn the language, and this knowledge will enable them to intervene into genes and gain mastery over the genetic program of humankind.

The Human Genome Project has shown that 99.5% of the genome is the same in any given person, thus 0.5% determines a difference between people and thus determines the genetic identity of an individual as a singularity. The majority of genetic material thus determines the species, the norm of the species, *normality*. The information provided by the Human Genome Project has helped to launch further research into the "normality of the genes." Similarly, as we have discussed so far, early observations usually result in general mapping or mapping of the norm, while further exploration is focused instead on looking for answers to particular questions, thus mapping becomes specialized and also individualized since every disease is individual, is a singularity. Where are we, then, with the genetic map at the present moment? Further investigation of the human genome has focused on determining the regular base pairs and establishing another map that would depict proper connections and abnormalities, such as those that determine Alzheimer's and other such diseases. This will be a map of a human as a prototype, particularly concerned with genetic normality in order to help us detect any abnormalities. The survey of the human genome is subjected to the next generation of gene therapy.

The field of genomics, especially human genomics, has been closely connected to the computer sciences from its very beginning. The display of the Human Genome Project results required significant bioinformatics resources, i.e. the application of computer sciences and information technology to biology and medicine. These joined fields of biology and computer sciences aim at understanding life as a program. At the moment we can determine that life is indeed such a program, and now that we have decoded it, we have the ability to intervene into and manipulate it. To study gene expressions, i.e. the organism's genotype and its phenotype, as functional genomics does, involves the same aim: to decode the language of genes or, perhaps, to develop short cuts in our ability to manipulate genes (without completely understanding them). Anyhow, one can see that the aim to understand is actually oriented towards imminent intervention. The scientific aspect, which we could understand as the collection and production of knowledge, is tightly linked to the engineering aspect of these strivings and doings.

1.4.2 Technical Images in Science as Means for Establishing the Truth

The regime of *perspicere* is about finding truth with our senses, particularly vision. Tools can be used, but it still comes down to the same thing: piercing through surfaces into the substance in order to dig up the truth, to see (even if through an opaqueness of membranes) what is actually there. This is understood as gaining knowledge about what it *truly* is: how it is structured and how it works. Descartes, however, did not trust the senses, although his work was concerned with getting the truth.

Experimental natural philosophy founded in the seventeenth century truly believed in *perspicere*, in the sense of the patient observation of the world and its structure, of gazing which gradually assures the truth. *Perspicere* is to be essentially understood as a visual regime because it necessarily involves the sense of vision. The contrary philosophical pole to empiricism developed a critique of sensuous experience and replaced it with a trust in rationality. The truth is not what I can see, but what I can think and comprehend. But does something change if we claim that the means of deception used in baroque visual arts, architecture, decorative arts, and literature were not truly about a distrust in senses, but rather quite the opposite, about the empowerment of artists who have become aware of their ability to master the means of visual appearance, the means of "visual writing" and of language? Art thus becomes the art of "making a sign," and this is the mastery of constructing a representation: of getting meaning into the picture.

Representation does not only or first of all originate from a visual skill or visual operation, but is basically conceptual. Representation is about mastery over the construction of semiological networks, among which one can include semantics. It is all about constructing meaning or rather *constructing reality*. For this reason, the relationship to truth changes. In the final instance, the truth becomes something that one in power can establish. Marvelous baroque ceilings, such as those of Pozzo or Pietro da Cortona, thus do not deceive the viewer. Instead, they enrich reality. If one were to claim that these are mere illusions, we would reply with Maurice Merleau-Ponty's critique of objectivity. In the case of the Müller-Lyer illusion involving two lines of equal length that we perceive as different in size, psychologists would say that we are wrong because they presume that there exists an objective word, and they thus claim that our perception is fallible. But how could we know what is real, and who is authorized to tell us what this real is? In his critique of Cartesianism, Merleau-Ponty would say that there are no two lines which are *objectively* the same, though we falsely estimate their lengths because of the directions of their arrow endings, the cause of the "optic deception." The alternative of equality or inequality is only possible in an objective world, but these lines are neither equal nor unequal. Each is perceived in its own context, as if they did not belong to the same world[103]: "We must not, therefore, wonder whether we really perceive a world, we must instead say: the world is what we perceive. In more general terms we must not wonder whether our self-evident truths are real truths, or whether, through some perversity inherent in our minds, that which is self-evident for us might not be illusory in relation to some truth in itself. For in so far as we talk about illusion, it is because we have identified illusions, and done so solely in the light of some perception which at the same time gave assurance of its own truth. It follows that doubt, or the fear of being mistaken, testifies as soon as it arises to our power of unmasking error, and that it could never finally tear us away from truth. We are in the realm of truth".[104]

---

[103]Maurice Merleau-Ponty, *Phenomenology of Perception* (London, New York: Routledge, 2005), p. 7.

[104]Ibid., p. xviii.

All of this is linked to an acknowledgment that a medium is not transparent, it is not a window through which we get a look at the truth out there. This actually did not first take place with the linguistic turn in the twentieth century. It was announced much earlier, either with representation in the seventeenth century or even with the Renaissance's constructions of landscapes on two-dimensional screens (or even earlier, as this technique was also mastered by the Ancient Romans). What happened with the linguistic turn and with linguistic philosophy in the twentieth century is instead the deepening of this epistemological crisis, this distrust in the mission and operation of searching for objective truth or the proper way to such a truth. This was recognized in hermeneutics and phenomenology, in linguistic philosophy, and most notably in structuralism and post-structuralism.

Flusser states that technical images "are not windows but rather images, i.e. surfaces that translate everything into states of things; like all images, they have a magical effect; and they entice those receiving them to project this undecoded magic onto the world out there."[105] Here he follows semiology. He summarizes what Roland Barthes claimed for photography. The medium of photography is particularly interesting since it seems to be the last fortress of the myth of medium transparency, which was clearly defeated with semiology. Barthes demonstrated that *the syntagm of the denoted message*, which is a message without a code (at least a conventional code), *naturalizes the system of the connoted message* (where there are cultural codes at work).[106] To make his theory clear, Barthes focuses on the advertising image, "because in advertising the signification of the image is undoubtedly intentional; the signifieds of the advertising message are formed *a priori* by certain attributes of the product and these signifieds have to be transmitted as clearly as possible. If the image contains signs, we can be sure that in advertising these signs are full, formed with a view to the optimum reading: the advertising image is *frank*, or at least emphatic."[107] The issue Barthes deals with is this: how does meaning get into the image if the image is re-presentation, a repeated presentation, i.e. a copy (because according to ancient etymology, the word "image" should be linked to the root *imitari*)? His formulation of the question is rhetorical: "can analogical representation (the 'copy') produce true systems of signs and not merely simple agglutinations of symbols?"[108] And this is how Flusser comprehends images: "Images are not 'denotative' (unambiguous) complexes of symbols (like numbers, for example) but 'connotative' (ambiguous) complexes of symbols: They provide space for interpretation."[109] And how is this linked to our reflections on the regimes of *perspicere* and *proicere*? We can see that for Flusser technical images are no windows, but images, therefore they do not support the regime of *perspicere*.

---

[105]Vilém Flusser, *Towards a Philosophy of Photography*, p. 16.

[106]Roland Barthes, "Rhetoric of the Image," in: *Image-Music-Text* (New York: Hill & Wang, 1964), p. 51.

[107]Ibid., p. 33.

[108]Ibid., p. 32.

[109]Flusser, *Towards a Philosophy of Photography*, p. 8.

Furthermore, Flusser links the concept of the image to that of projection. Images are abstractions of "something out there," which they make comprehensible to us—we could even call this process codification. Flusser writes that the "specific ability to abstract surfaces out of space and time and to project them back into space and time is what is known as 'imagination.'"[110] Therefore, "[t]he world is our projection."[111] Imagination is the precondition for the production and decoding of images or "the ability to encode phenomena into two-dimensional symbols and to read these symbols."[112] Production and particularly the distribution of images does not mean revealing the truth and opening up surfaces to enable insights into the deepness of this truth, but rather projecting the connotations and complexes of symbols that are not even decoded anymore: "Human beings cease to decode the images and instead project them, still encoded, into the world 'out there' ... Human beings forget they created the images in order to orientate themselves in the world. Since they are no longer able to decode them, their lives become a function of their own images: Imagination has turned into hallucination."[113] This is how we have come into the service or function of our own products.

There is a crucial difference between traditional and technical images, according to Flusser: "Ontologically, traditional images are abstractions of the first order insofar as they abstract from the concrete world while technical images are abstractions of the third order: They abstract from texts which abstract from traditional images which themselves abstract from the concrete world. ... Ontologically, traditional images signify phenomena whereas technical images signify concepts."[114] This comprehension could be linked to the epistemes we were discussing before: traditional images were linked to the world out there, they were resemblances; they thus belong to the Renaissance episteme according to Foucault. By contrast, technical images are representations; they build upon the project that started with the classical episteme. They don't signify things out there; they are thus not transparent windows, but rather representations of concepts. They build a conventional communication system, as in language. However, there is a difference between language and technical images for Flusser: "Technical images are difficult to decode ... their significance is reflected on their surface: just like fingerprints, where the significance (the finger) is the cause and the image (copy) is the consequence."[115] Technical images are tricky because they seem to resemble or copy the world out there, but they are in fact encoded meanings. Therefore, Flusser notices a magical functionality within them. However, this was detected earlier by Walter Benjamin in his recognition that photography and film are *loci* for ideology: "The function of technical images is to liberate their receivers by magic from the

---

[110]Ibid.

[111]Flusser, *Writings*, p. 90.

[112]Flusser, *Towards a Philosophy of Photography*, p. 8.

[113]Ibid., p. 10.

[114]Ibid., p. 14.

[115]Ibid.

necessity of thinking conceptually, at the same time replacing historical consciousness with a second-order magical consciousness and replacing the ability to
think conceptually with a second-order imagination."[116] Technical images displace
texts. Thus, the invention of photography is for Flusser a historical event as equally
decisive as the invention of writing. If history has been a struggle against idolatry,
then with photography post-history begins as a struggle against textolatry. This is
how Flusser speaks about the visual turn.

With the visual turn, technical images are now projected onto the world out
there, although the concepts or ideologies involved are no longer under our control;
they are produced and distributed faster than they can be comprehended, but they
still function, ideology still gets to work. Flusser has noticed textolatry in
Christianity and Marxism: "Texts are then projected into the world out there, and
the world is experienced, known and evaluated as a function of these texts."[117] But
in worshiping texts, or rather distributing the concepts encoded in them with the
distribution of the text itself, projection was still rational, intentional, organized,
whereas in the case of new media projection we have now become functions of
apparatuses.

What can we conclude regarding today's trust in the regime of *perspicere*? It has
been subjected to a critique. However, trust in it is actually still very much alive in
the techniques of the repressive apparatus of the state, which relies on every sign
that holds an indexical link to the object, thus a link that supposedly testifies to its
truth: photographing, fingerprinting, genetic profiling. Furthermore, this regime is
still at work in biology (microscopy), in astrophysics (telescopy), in medicine (all of
the body imagery technologies enabling us to see what cannot and could not be
observed with bare vision or the naked eye). We are now gazing through opaque
membranes, we have developed focal views, a variety of lenses that enable us to see
what exists but is not visible by itself, both micro- and nano-scale worlds and the
very distant worlds of other galaxies altogether. We can see deep into the universe,
meaning far into space and back in time. Vision is in these cases directly linked to
the assurance that knowledge crucially depends on vision. And comprehension gets
deeper, more complex and also more "*truthful*" when seeing goes deeper and
farther, when it becomes more focused and selective. Developing the tools to
improve or enhance *perspicere* enables us to get even closer to the truth.

Do we think or see first? How do we know things? Actually, the Kantian
epistemological question as to how is science possible is still open. This doubt is
significant for the whole of modernity. Exactly in this issue lies the attractiveness of
discussing the two regimes (*perspicere* and *proicere*) as working for and against
each other. The human species has the desire to rule the world, and Flusser was
aware of this ambition. But he also knew: "In order to do that, we first have to

---

[116]Ibid., p. 17.
[117]Ibid., p. 12.

perceive it."[118] And what does it mean to perceive? "To perceive means to assimilate thinking to extension."[119]

In his essays on discovery published in *Artforum* in 1987 and 1988, Flusser pays attention to the rising field of biotechnology. Discussing the research processes and epistemological issues, starting with the concepts of the quantum theory, he presupposes we live in the time of uncertainty. He suggests that perhaps an unfixed perspective is a kind of better knowledge with which to think, a better space in which to live. In the first essay from the series, we find his thoughts on how to define new knowledge: "Any new knowledge that might have been gained is hidden in a gray zone of assumptions."[120] He offers a hypothetical case in which scientists strive to discover the origin of Earth's life. We have explorers from genetics return from an expedition and announce they've found some new genes, while explorers from molecular physics return with the information they've found regarding some new molecules. If we were to translate this example into our previous discussion, the explorers are the scientists, digging the holes, piercing through the surface, the obvious, that which is visible at first glance, striving to reach deeper and finding more, finding new things, getting surprised, gaining new comprehension, arriving at understanding, coming closer to the truth. However, the *perspicere* that is at work here is not only a disinterested opening up of opaqueness, but it is also fundamentally determined by what we already possess, by what we already know. *Perspicere* is combined with *proicere*. One way of putting this would be to say that we use *perspicere* to discover and then rely on *proicere* to project ideas. Or we could say: we use *proicere* in undertaking *perspicere*.

1.4.3 Creativity of Biotechnology

It is because of their smallness that amino-acids were discovered so late: "And once they were discovered, however, it became possible to manipulate them."[121] Flusser believes that biotechnics "is a discipline out of which a whole world of artificial living beings—living artworks—will arise."[122] Manipulating biological information in the sense of the production of new information by recombining the elements of information already available to us is what Flusser calls "variational" creation, whereas it is something else when there is a new element (some noise) added to the production of new information, which Flusser calls "true" creation. Flusser notes in 1988 that if biotechnology has so far restricted itself to variational creation in

---

[118]Flusser, *Writings*, p. 88.

[119]Ibid.

[120]Vilém Flusser, "On Discovery," in *Artforum*, New York, Vol. 26, No. 1 (September 1987), p. 10.

[121]Vilém Flusser, "On Discovery," in: *Artforum*, New York, Vol. 27, No. 7 (March 1988), p. 14.

[122]Ibid.

combining the elements of available genetic information, then God used the other method (true creation), and there is no reason why biotechnology should not do the same at some point in the future.

Flusser then takes into consideration the concept of morphogenesis (the birth of form) to discuss the issue of creativity, and he concludes that there are two principal ways of creating new form. First, several old forms can be combined into a new one —this method was known in the ancient Greek *chimera*, made up of the combined forms of a goat, lion, and serpent. Today we have the "geep," a genetically engineered derivation of the goat and the sheep, which is a contemporary version of the same principle. But if something new is added to an old form, as in a chess game one might introduce an elephant, the game would be completely changed.[123] We shall discuss this question, whether biotechnology has gone so far as to become godlike. There is one particular concept that will play a significant role in our consideration of this.

Photography as a medium, seemingly transparent, has been subjected to criticism with semiotics. However, photography still has a connection to the scene "out there," just as the television news does. What we were emphasizing earlier is that these are all conventional discourses, ones that are well organized; there are strategies of containment (to borrow John Fiske's term) at work. We have thus been pointing out the importance of constructiveness. Marshall McLuhan recognized cubism as the art movement that clearly performed the critique of media transparency, demonstrating that the "medium is the message." Analytical cubism clearly displayed the process of analyzing a scene that was taking place during the process of painting, thus the painting was already very much about painting. But it was *synthetic cubism* which truly ceased to be about the world "out there" and became instead the occasion for mastering the world by mastering painting itself. In terms of what we have discussed so far, we could say that synthetic cubism broke with the idea of the transparency of painting (if it was for any reason still alive at that time) as clearly as possible and manifested the constructive nature of painting. But, as regards our interests here, let's pay a bit more attention to this concept of synthesis.

Etymologically, the term originates from Gr. *synthesis* meaning composition, from *syntithenai* meaning put together, combine (*syn*: together and *tithenai*: put, place). In seventeenth-century Lat. *synthesis* means composition, set, collection, and in the nineteenth century *synthetic* refers to products or materials made artificially by chemical synthesis (hence artificial). As the example of synthetic cubism demonstrates, synthesis is not *perspicere* (looking through, into). It is instead all about dominating the medium, and thus the world, by taking particles from the world and constructing a new composition. It is the technique of synthesis which is at work in photo-collage, film montage, and in assemblage. And it is the latter that presents the model of the rhizome for Deleuze and Guattari. Synthesis brings together diverse elements; it builds a world of heterogeneity. The baroque and

---

[123]Flusser, "On Discovery," Summer 1988, p. 17–18.

rococo's visual and decorative arts operated with synthesis in order to form new realities, new worlds, new living beings, chimeras. And it is exactly the technique of synthesis that has become crucial in biology as it transforms biology from a science into a technology. Biology has become another field of engineering, one that engineers living structures.

Synthetic biology is a recently enthroned field of knowledge-engineering, one that applies computing to biology.[124] The leading researchers in the field do not necessarily start out in the field of biology; they can even be computer scientists, and this attests to the transdisciplinarity of the field: it joins together biology, the computer sciences, chemistry and various technologies. For Ron Weiss (MIT), one of the founders of the field, the idea of synthetic biology is to glue together DNA parts one can even get online. The biological parts of the DNA sequences of defined structure and function that are designed to be incorporated into living cells are called BioBricks. They are used to assemble bio-circuits. These biological circuits program biological machines. They can order programmatic commands, for example by making a protein that creates the color blue. One of the prospects of synthetic biology in medical application is the repairing of human tissue with the use of bacteria. In medicine synthetic biology promises imminent solutions with the technique of disease targeting, which would function in this way: if a cancer cell is encountered, a protein is made that kills the cancer cell; if not, it just goes away. There are fears that the body would recognize the genetically modified cell as a non-self and react, for example, by rejecting the protein or forming a new type of cancer. To avoid this, synthetic engineers see two solutions: 1. use only proteins that come from the body—this could then make an interesting contribution to the paradigm of regenerative medicine we discuss in the final chapter—or 2. build molecular computers that are able to build the function into the RNA so that the body would recognize the protein as part of the self. MIT has built targeting technology that works in vitro, but has not yet applied it to a human body. Researchers from Stanford University have been focusing on developing a molecular computer that could control the body's immune response with molecular controllers that enforce the survival proliferation of T cells (T lymphocytes), which is a contrary process to apoptosis. They believe that new generation therapies will use this type of strategy. NASA has developed a method of targeting as well: synthetic organisms are to be put into the body to treat astronauts from radiation. The idea is to use engineered bacteria and combine this technique with nano-technology: a bio-capsule composed of carbon nano-tubes will respond to radiation and release therapeutic molecules. This means we can expect a new sort of cyborg. At the intersection of electrical engineering, a new field is also emerging: synthetic neuro-biology. At the MIT electronics laboratories (Ed Boyden), they are

---

[124]The term synthetic biology was actually introduced a century ago (Stéphane Leducs, 1910). In the 1970s the field became especially promising: Waclaw Szybalski was aware that it actually had unlimited expansion potential, particularly in the development of new control elements and application of these elements to existing genomes, to say nothing of the construction of whole new genomes altogether.

working on developing new kinds of computers and engineering the most complex computer of all: the brain. They are trying to control electrical pulses so they can enter information into them in a manner similar to how you enter information into a computer circuit. They are using illuminators (lasers) to do this. Using algae's ability to photosynthesize, light pulses are converted into electricity, and proteins hit by light generate electrical pulses that control the neurons. This introduction of control of the brain has been tested on a mouse, but not yet applied to a human. The technology seems promising for treating Alzheimer's disease.

The field of synthetic biology is actually not much different from genetic engineering—both involve genetic programming, which means encoding functions within DNA. Synthetic biology thus sees a challenge in figuring out how to do precisely this. It is no wonder that one of the central figures in the field of synthetic biology was also key to the project of human genome sequencing.[125] Since 2005 Craig Venter has been intensively involved in synthetic genomics. He believes that genomics has the power to radically change healthcare and economy (he is developing the next-generation biofuels with synthetically modified microorganisms). Venter is currently aiming to create a life form (*Mycroplasma laboratorium*). In 2010 his team announced in *Science* that they had created "Synthia," a kind of bacteria that had hitherto never existed in nature. They successfully synthesized the genome of the bacterium *Mycoplasma mycoides* from a computer record, and then this synthesized genome was transplanted into a cell of *Mycoplasma capricolum* bacterium, from which the DNA had been removed. In other words, a long DNA molecule containing an entire bacterium genome was plugged into a computer, where it was manipulated as software, then Venter's team extracted and discarded the DNA from a similar cell and finally introduced the created DNA into the emptied cell. The parent of "Synthia" is a computer, and it came into physical existence as a DNA print. The boundaries between the computed and biological have literally blurred. This creation has thus been referred to as "synthetic life." However, despite the success of the already heavily customized genomics and genomes, "[t]here are great challenges ahead before genetic engineers can mix, match, and fully design an organism's genome from scratch,"[126] Paul Keim has noted. Among the prospects for the field of synthetic biology, perhaps the most radical perspective comes from so-called re-writers, who believe that natural biological systems are so complicated that we should rebuild them from the ground up in such a manner that we can thereby provide engineered surrogates that we better

---

[125]Craig Venter was the leader of the team in Celera that, parallel with the public Human Genome Project, worked on generating the sequence of the human genome. Both teams announced the mapping of the human genome at the same time: Celera published its results in *Science*, and the group managed by Francis Collins of the National Institute of Health in the U.S.A. followed with publication in *Nature*. Celera used DNA from five demographically different individuals, one being Venter himself. The teams, however, used different methods: Celera's method was shotgun sequencing, while the Human Genome Project used the clone-by-clone method.

[126]Elizabeth Pennisi, "Synthetic Genome Brings New Life to Bacterium," *Science* 21 May 2010: Vol. 328 no. 5981 pp. 958–959. http://www.sciencemag.org/content/328/5981/958.full, 7-5-2012.

understand and with which we can more easily interact. This idea comes from the computer sciences, where code refactoring is a technique used to restructure a code by altering its internal structure without changing its external behavior, in order to improve the code's readability and reduce its complexity for the enhancement of the source code, as well as to improve its extensibility.

Flusser noticed that "biology has stopped being a natural science and has turned into genetic engineering".[127] Is it still possible to call these engineering practices science, if science means the exploration of the world, disclosure of truth, and production of knowledge? We are witnessing an epistemological rupture. Science has become knowledge-technology, where knowledge is produced for the sake of getting used for intervention that will in the final instance assure certain profit (or loss).

Flusser has discussed the issue of "becoming godlike" in regard to biotechnics. If a form were created that had never existed before, this would be an instance of true creation. And in the case of true creation, we would be dealing with magic and a magical power characteristic of artistic creation. This bringing to life will result in something its creator will be incapable of understanding. We could say that Flusser claims something similar here for technical images, which are projected into space without being decoded. True creation is the province of a genius: "we might suppose that an artist—or a genetic engineer, or any kind of 'creator' for that matter [i.e. living matter]—is the more godlike the more he or she has access to it."[128] The model of variational creation is evolution. It is used not only in computing, but also in genetic engineering (including synthetic biology) since it produces combinations and permutations of already existing elements of information. By contrast, true creation would challenge the very idea of evolution, which is exactly the aim of present-day life-engineering. Perhaps Flusser's distinction between the two simply has to be discarded. Cloned mammals like Dolly (the first mammal clone, born in July 1996) were produced by somatic nuclear transfer: the nucleus of the skin cell of the genetic parent (a grown up sheep) was inserted into an emptied egg of the egg-donor sheep, and the artificially "created" zygote was grown in a laboratory (of the 277 udder cells used, 29 grew into embryos, but only one embryo developed into a fully formed lamb). The clone is genetically identical to the nucleus donor. It is not only possible to "create" a clone from only one donor, donating the egg and the nucleus, but also to generate the egg cell from a stem cell, thus having the same donor be a male donating only adipose tissue. Still, genetically modified and synthesized organisms are not *created* in the sense of bringing something formed from dead matter into life, which was the original, Biblical sense of creation: "And the LORD God formed man of the dust of the ground, and breathed into his nostrils the breath of life; and man became a living soul."[129] This Cartesian dualism between dead matter and a living soul is present in Venter's aim to *create* a

---

[127]Vilém Flusser, "On Discovery," in: *Artforum*, New York, Vol. 26, No. 2 (October 1987), p. 12.
[128]Flusser, "On Discovery," Summer 1988, p. 18.
[129]*Genesis* 2:7.

life-form originating from another, non-living source (as for example from a computer program). We would claim that the achievements of biotechnology in the manipulation of life do not simply re-present this ancient dualism between body and soul. The boundaries between living and non-living matter have become blurred, and the concept of life has acquired novel dimensions that are quite incomparable with traditional ones. The notion of having life "on" and "off" is simply obsolete, since life can get dissolved, dispersed, diluted, or delayed.

The ambition to "create life" is either one of "playing God" or of *mimesis* in the Aristotelian sense, i.e. to resemble the performance of nature, this time in its ulti-mate role in the origin of life. The evolution of living creatures after their creation could be interpreted as a realization of God's wish, with living beings having developed according to a program. Recently, the human species has acquired self-confidence in understanding this program well enough to be able to intervene into it, to change it or even to apply its own program with engineering methods. All sorts of living structures have become decisively dependent upon the program applied to the whole of this living world by man. We are facing a new moment in the transformation of biopower, which will be discussed in the last chapter of this book.

According to Flusser, variational creation is a method requiring a lot of work being done not just with computers, but also with biotechnology: "biotechnics is doing the same thing natural evolution does—variational creativity, the sole dif-ference being that it does its work not by chance but according to a deliberate program."[130] Variational creation operates within given possibilities, one could say within the "natural apparatus", similar to how Flusser once considered pictures produced within the apparatus of photography, how every particular realization within this program exists as a potential, even if it will never actually be realized: "Every shape in which Earth's living beings could manifest themselves is encoded within the existing genetic information as a potential, a virtuality."[131] If Flusser introduced the same logic as he used in *Towards Philosophy of Photography*, then he must have appreciated the resistance to the program that is to be performed by the creative agents.

If we accept Flusser's notion of true creation as the creation of living forms that have never existed before, then many products could be regarded as true creations. However, we could make out an original interpretation of the concept of creation. With adding something completely new to the existing form, for instance putting sight into wheat, one would produce true creation, however in the language of modern communication theory adding something new to the already existing form would mean adding *noise*.[132] In informatics noise would be the unwanted com-ponent of communication process, since it is an unintended element disturbing the transparent communication of the message. Thus, a simple conclusion would be

---

[130]Flusser, "On Discovery," March 1988, pp. 14–15.

[131]Flusser, "On Discovery," Summer 1988, p. 18.

[132]Ibid.

that signal has to be purified from noise. But if one is recognizing certain forms of communication as intended communication at work in the consolidation of dominant ideologies, then noise is to be welcomed in such communication. Such is the concept of the forces of disruption in television discourse offered by John Fiske,[133] which are actually noise bothering the communication of the dominant or intended message of the television news. Forces of disruption are unintended elements in the constructed discourse, which are available to the proper, we could say critical reader. In a similar sense, art can be understood as noise added to the dominant discourses. In this context adding noise would mean making a disturbance in the program, *resisting the apparatus of power*. According to such an understanding, art does not *create* so much as it *resists*. If we recall Flusser, the notion of the true creation means adding noise to the program of nature. Therefore, true creating is a case of resistance. Creativity is thus in resisting the prevailing program. This act, resisting, is in this case however not being done by artists, but by biotechnologists. Yet, biotechnology itself produces deliberate programs. This takes place exactly because of the projection of the theory is directing the acts of intervention, because it is not an observing, but an engineering practice.

In this sense biotechnology is adding noise to the communication of nature. But who is adding noise to the communication of biotechnology? Shall we claim that biotechnology is still mostly concerned with the purification of the signal, as it is working hard to arrive at a functional outcome, a living product expressing programmed features? In biotechnological practice, in spite of all the efforts to engineer the product perfectly, it is likely that it will not turn out as conceptualized and that several unplanned features will be present as well in the final result or application. This is the reason to hesitate over the application of these technologies to human beings. And as happened with Dolly, she was born old and died sooner than anticipated. So doesn't the apparatus of biotechnology need added noise, more disturbances in its discourse, because it is itself rather heavily troubled? Flusser aimed to contribute a defense of creativity; true creativity does not take place within the framework of the apparatus or according to a plan, but instead ensures surprises as it intervenes into existing processes taking place according to their program. Creativity is therefore a subversive act. For Flusser, biotechnologist and artist seem to have become one and the same person. Resistance is directed towards nature. But if biotechnology is itself an apparatus, who is resisting it? Where is the creator who resists fitting into the machinery that produces variations of the program of biotechnology?

"Who will be the Disney of the future," Flusser asks and answers: "He or she might, I suggest, be a molecular biologist."[134] Disney not only paints the world in his colors, as does the Pink Panther, but he is also the one that holds the pencil, who organizes the whole of the work from concept to final realization: Disney is an *engineer*. But what is the role of the engineer within the apparatus of

---

[133]John Fiske, *Television Culture* (London, New York: Routledge, 1987), p. 281–308.
[134]Vilém Flusser, "On Discovery," in: *Artforum*, New York, Vol. 27, No. 2 (October 1988), p. 9.

biotechnology? According to Flusser the engineer has the potential to be a true
creator; he might be able to play God, thus acting as the subversive element in the
game of nature. Etymologically, the term engineer appears in the early fourteenth
century and means the constructor of military engines; it originates from the Old
French *engigneor*, from Lat. *ingeniare*, and the term *ingenium* means inborn
qualities, talent. The term is linked to the term *engine*, a mechanical device, but also
skill, craft, from the Old French term from the twelfth century *engin* meaning skill,
cleverness, but also trick, deceit, stratagem; war machine.

Chapter 2
Collection of Identity: Body Prints and Identification

Abstract In the second chapter genetic prints are examined as representations in relation to what they represent. In science and repressive apparatus of the state body prints are considered to be authentic indexical signs, while in some recent art projects the genetic language and the issue of identification are challenged. The relations between the signifying structures and the body are analyzed in this chapter. Collecting body prints and the issue of identification are comprehended as strategies for biopower to be profoundly exercised over individual bodies and over populations (biopolitics).

We live in an age in which societies are accelerating the development of technologies that track individuals in space and time. There are several ways of mapping identity that take place at the same time and in different manners. Contrary to the mapping of space through the body (to grasp the memory of one's experiences), which so far has been performed only for research purposes, there is another mapping that is taking place in the "real" world, in the world in which we all live. It is a mapping that is putting one into other maps, of not only cities, airports, and countries, but also those of internet locations and social networks of all kinds. In the previous chapter we discussed several attempts to map the human body in order to map "the human"; in this chapter attention is instead devoted to mapping the individual through the body prints collected in social or political maps. At the current moment there are geographical areas in which surveillance technology is already collecting so many location points of our bodies in space that for each of us there is a file with enough data to establish a personal map of our movements in that area in a certain period of time. Not only is the life of each individual being tracked and stored so that any sequence can be reconstituted in a moment and then analyzed, but also there are patterns being established from our repeated actions, trajectories, and meetings so that it is possible to calculate and draw probable future personal maps and establish possible connections with other personal maps. It is not our aim here to discuss how this information could be used or "misused" for the analysis and prediction of our behavior in the service of economic, political, intelligence, military or religious goals, nor do we intend to discuss what drives

© The Author(s) 2017
P. Tratnik, *Conquest of Body*, SpringerBriefs in Philosophy,
DOI 10.1007/978-3-319-57324-3_2

power to do such things. We are interested instead in the fact that our bodies are being topo-graphed and chrono-graphed, and particularly in the fact that this is being achieved by processes of mediation and with techniques of body printing and body graphing. The burgeoning use of such technology to collect our body data and to map it establishes the very foundations for an unprecedented exercise of bio-power over individual bodies and mass populations.

In order to establish such maps, various sorts of body representations need to be collected, since representations are what assure information about the presence of a particular person at a particular location, the data which will then put that person on a map of a particular area. The technologies used have to be as precise as possible, they have to store as many details as possible and also collect various sorts of data: the face will not be scanned, digitalized and then recognized from one perspective only; perspectives will instead be combined to establish a three-dimensional simulation of the head, visual recordings of physiognomy will be supplemented by iris prints, etc. In this particular context, the most important aspect of such body-graphy is the basic requirement that needs to be fulfilled in order for this mapping technology to fulfill its function: representation has to be *caused* by the body, there has to be a trace of the body, the representation has to be body's *print* in order to serve as proof of bodily existence at that particular location and at that particular time. In this sense one could say that the representation has to be reliable and infallibly determine that one particular individual was present at that particular train station at the time of the bombing. In this sense the technologies of body printing have a particular function: *to establish the presence of a person with certainty*. Our aim here is thus to discuss the infallibility of these body prints. The development of technologies that will assure the greater amount of detail in a scan, the development of new technologies for body printing, the increase of surveillance locations: each of these factors attests to the fact that present technology must be insufficient, otherwise it would not need any improvements.

The other aspect that we want to discuss and that is connected with the issue of the relationship of the body print to the body itself is the issue of *identity*. It is not only important that prints verify the presence of a body in time and space, they also need to testify to the fact that the body at issue was *this* particular body, a singularity. The prints serve as *body identifiers*. This seems to be one and the same issue: to put one particular individual at a particular location at a certain time. It holds to a certain degree. But the issue of identity is worth discussing with special attention. The prints that we will analyze are not just any prints, but rather traces of an individual, such that individuality is determined exactly by means of *identification*.

If one takes this perspective, she might be surprised to find out the significance that various kinds of marks have for establishing one's identity, not only for specialists, such as forensic pathologists or doctors, but also for those in her social life and even for herself, since she too is constantly using signifiers. The identity of the dead is collected through marks; for the sake of our colleagues, we aim to build and exhibit signifiers that communicate the signified we intend; for bureaucrats we keep special marks in our pockets that they will comprehend as signifiers of our identity, etc. We are constantly collecting marks that encompass an identity—not so as to come closer to this identity, as is usually said to be the case, but rather to establish it, to collect it altogether. And taking into account the extent to which this is a process, we see that we are in a constant process of becoming, of becoming our identity. Structural psychoanalysts have acknowledged this process of collecting the marks through which we strive to attain knowledge about ourselves, about who we are, about what our identity is. But it is another question entirely whether this process is ever completed, whether identity can ever be collected or established for certain.

So far we have discussed some aspects of the process and function of re-presentation as regards body imagery, but we have not yet emphasized the role of body representations for the biopower of the state. We have seen how much scientific discourse and medical practice rely upon the credibility of body imagery. Similarly, the repressive apparatus of the state strongly relies upon the credibility of technology to produce representations that can be understood as absolute representatives, thus proving the identity of the body they represent. In such a sense, these representations have to enclose the essence of the thing. If they were "merely" imitations, then there would be no distinction between them and painted images. We could thus identify a person depicted on a painting—as a matter fact, if Leonardo da Vinci is to be believed, the painted portrait is the best possible representation; in that case the person depicted would be even more identical to herself than she could be for someone deducing her identity from her "mere" presence. According to Renaissance belief, the painter's mediation assures a surplus, he adds something that the natural appearance itself does not possess. He has to dig into the essence of her identity and bring it up to the surface of visibility in order to represent it in the medium of a painting. The point is that this is an essence that does not appear naturally in the live presence of the person itself. It is the painter who can reveal it and display it. The mediation of the painter is crucial in getting a quality result. The essence of the person (her identity) is radiated in the depiction. However, the painting does not bear physical contact with the person portrayed, it is not *her* print, the trace of *her* presence. There is no proof, in fact, that the person ever even existed, let alone any certainty that she is *identical to herself* in this painting. Nonetheless Mona Lisa is *identical to the person she represents*, she is *the same* Mona Lisa. Because the portrayed Mona Lisa represents the idea, the concept, the existence of the person depicted in this painting in the real life is not relevant. The representation is an autonomous reality.

Painting is a constructed discourse as television news are. Painting is not a print, it is an iconic sign. The repressive apparatus of the state however requires indexical signs, i.e. signs that are physically caused by the objects. Body prints are nevertheless signs. As mediations that supposedly guarantee identity they are not to be questioned, because they are not understood as mediations. They enjoy a dogmatic status instead, one equivalent to that of myths. Photographic portraits, fingerprints, iris prints, DNA profiling, etc.—we will treat these body prints as mediations which in our political reality function as a person's identifiers. Here we pay particular attention to the question of mediation in those biotechnological representations that enjoy the reputation of assuring a certain, infallible identification.

In forensic science, DNA fingerprints serve as the ultimate body print for tying an individual to a crime scene. Can this print be deconstructed, though? Paul Vanouse has systematically tried to question the authority of DNA profiling. In *Latent Figure Protocol* (2007), he has created visual "images" by using known DNA patterns—with the known visual outcomes that are formed in electrophoresis by the DNA of bacteria, he has put together new visual compositions, which in simple digital visual designs (consisting of light and dark spots on a quasi-pixilated surface) represent visual motifs (such as: chicken and egg, the pirate symbol of a skull with bones, ID, 01 and the symbol ©). The "representation" or final visual outcome, which on the basis of similarity represents something (for example a chicken), has nothing in common here with the singular material, that is to say the biological samples whose DNA profiles constitute the visual representation. Material here is used in keeping with the same principle a painter applies when using colors: for building a visual surface that is not motivated by its constituents, its material elements, but rather by the optical similarity of that which is painted to that which is portrayed. If the picture were motivated by its material, then red paint could not become the color of the precious cardinal's garment, and painting would not have become the art of transubstantiation; instead a painting made out of organic pigment would become a field of massacre, and to paint would mean to massacre. The latter is exactly what Jan Fabre has done with his *Heaven of Delight* (for which he embellished the ceiling and other elements of the Royal Palace in Brussels with 1.6 million bugs of the unprotected species *Buprestidae*). This work should be "beautiful" because of the strong aesthetic effects (with colors that change with light) of the decorated ceiling of the Salon de Glaces in the most prominent Dutch building from the 18th century. Yet, the unconcealed use of animals, millions of dead corpses, for aesthetic purposes produces perverseness of this act of design. The fact is that organic pigments have become awfully rare today and the question of motivated aesthetics constituents has also become almost superfluous because of the "inauthentic" origin of the materials usually used. For pigments available on the market a demand for substantial transparency of the material's origin is rather absurd because of the too complex mediation or chemical hybridization. Sculpture is closer to the principle of the motivated handling of a substance, which is one of the bases for the closeness of statuary art to conceptualism, insofar as the *David* is with and for Michelangelo the *David-in-marble*: it is an image of David and it is marble; it is a person evoking life and yet always dead and cold matter. It is this paradoxical duality encompassed in this unique entity that makes *David* appealing. The statue is not a deception—the marble is there, so it does not strive to be something which it is not—a living human being, for instance. In ready-mades the origin doesn't transform, and the artifactual ready-made does not represent anything else but itself; it stands exactly as such or *it is itself alone*, it is what it is supposed to represent: the removal of the principles of substitution and reference, the equalizing of the sign with the referent assure the conditions for the instant transition from the art of transformation to the art of thought conceptualization. Duchamp's urinal, despite putting into force the demand for artistic artifactuality and standing as a candidate for the change into a fountain, has not changed its shape in terms of form.

As a result, it is still a urinal, though in its essence this is not the case anymore because what Duchamp's artistic gesture accomplishes conceptually is the *trans-essentiation* of urinal into a work of art and into a fountain, and not a *transformation* or *transubstantiation* of something into something other than itself.[1]

The original postscript to the art tradition of transubstantiation, which in fact accompanies the painting of the whole modern era (or the majority of painting from this period), contributes to the contemporary advocacy of painting as alchemy, as in the work of Sigmar Polke. The magical practice of transubstantiation represents the peak of the tradition of transforming materials, as well as the return to a time before art in the Middle Ages, as far back as Greek antiquity, when *chymeia* denoted occupation with alloyed materials (the technological blending of metallurgy and color techniques), and even back to non-European cultures, especially the Arab-Islamic golden age, when *al-kimyá* meant the coexistence of experiment and treatise (theoretical discussions). At the vanishing point of the European medieval program of alchemy, which searched for a way from physical (sensuous) experience to the metaphysical gaze, the contemporary archeologist Siegfried Zielinski recognizes the idea of projection.[2] According to Zielinski, *proicere* is essentially about a magical tradition whose genealogical roots we find in pre-Socratic philosophy, while *perspicere* supports the logic of a penetrating view through surfaces that was established by the beginners of modern sciences: Johann Kepler (*Dioptrics*), Galileo Galilei, René Descartes, and Isaac Newton, whose big accomplishment for the natural sciences of the 17th century (the "physics of visibility") was very interested in the problem of transparency (seeing through), while catoptrics were mostly interested in *proicere* (Ger. *Aufsicht*—control, view from above) or in the optics of mirrors and reflections. *Perspicere* (seeing through reality, as enlightenment) was supported in science by the development of optical technologies: microscope, telescope, and other such devices.

Vanouse is using biological material in the manner of transubstantiation so that the produced representations do not have a substantial relationship to the object they represent. The artist just demonstrates the fact of mediation. He addresses the issue of constrictiveness of DNA depictions. With *Suspect Inversion Center* (2011)

---

[1]However, the signifier is not equalized with the signified here, which would become the referent itself; the semiologic system is in this case more complicated. In the place of the signified "urinal," what gets interpolated are at least two other signifieds: (1) the fountain as an architectonically designed and plastically decorated well and (2) the work of art. Besides, the functions of substitution and reference, which constitute the conditions for semiologic order, are in indirect connection with materiality here—materiality is subordinated to a greater totality which is the *thingness* of the urinal, it is the ceramic in its extensiveness, in its form, and it is also a ceramic with a function. Therefore, this *reality as thingness* also includes tool-being (as Heidegger analyzed in "Der Ursprung des Kunstwerkes," 1935). But the tool-being is in this case being subverted as a urinal; in its essence it is not a urinal anymore here, therefore the major event of this work of art is the process of *trans-essentiation*, while the process of transformation that would mean the changing of the form does not take place, nor does the process of transubstantiation, which would entail a changing of substance.

[2]Zielinski, *Entwerfen und Entbergen*, p. 19.

Vanouse refers to the controversy involving O.J. Simpson's encounter with DNA profiling.[3] He carries out a unique deconstruction of the DNA sign: the genetic profile of O.J. Simpson, visualized on electrophoresis gel, is constructed from the artist's own biological material during the time of the exhibition by Vanouse and his assistant.[4] The artist explains his artistic act: "Just as the term DNA finger-printing has led to misinformation regarding its efficacy, DNA evidence is also hyperbolized in popular culture. The 'CSI effect' is a reference to the phenomenon of shows such as the CSI franchise overstating the accuracy of forensic techniques, and exaggerating the abilities of forensic science. *SIC [Suspect Inversion Center]* is designed to creatively counter these mass-media dramatizations that typically misinform the public, and in so doing, provide audiences with the conceptual tools to understand current issues surrounding use of DNA imaging and databasing."[5] The question of the relation of the sign to its object and interpretant is re-opened in regard to very real issues of identification. For a lay observer, this possibility is surprising since genetic "fingerprints" are socially understood as body traces, as *indexical signs* if we may resort to Peircean terminology.[6] The index has a crucial and direct relationship with the object which it *represents* and for which it stands; it cannot exist without it as it is its cause, its direct inducer, just as fire is the inducer of smoke or an injury is a cause of pain or a footprint the trace of a foot. In a similar fashion, a DNA profile is supposed to be the trace of a body, its representative. The other word for this technique is DNA blueprinting, a term we find more suitable. Blueprint still expresses body printing while also attesting to the mapping nature of the technique, rather than just claiming it is about the direct physical imprinting of the body into the jelly substance.

The projects performed by Vanouse point out the interposition between the body and the display of the DNA profile, i.e. the *mediation* of the biological material or DNA profile as a sign. Mediation makes space for manipulation and thus rebuts the function of proving the truth. As is the case with any medium, this one can be *deconstructed* as well, and it can be shown how it was built and how it functions; we can also *construct* it from the constituents of some other disintegrated whole, as

---

[3]Simpson was a famous athlete who allegedly murdered his wife and her lover; the main proof was provided by the results of the DNA analysis of the biological samples which put Simpson at the place of the crime. In the room where the double murder happened and in a nearby hallway, investigators found biological material, and its DNA analysis showed proof of identity with the biological sample of Simpson. However, all the criminal charges were dropped against the celebrity when his legal representatives showed a reasonable doubt as to whether the material on the scene was indeed present at the time of the crime, above all on the basis of finding that the same material contained an additional substance that acted as an expedient against the coagulation of blood, which was not present in the blood sample taken from the celebrity and that created doubt about when the sample was brought to the crime scene.

[4]First presented together with two other projects at an individual show titled *Fingerprints* ... 27th of January and 26th of March in Schering Stiftung, Unter den Linden 32–34, Berlin.

[5]See *Suspect Inversion Center*: http://www.paulvanouse.com/sic.html, 02-28-2017.

[6]Charles Sanders Peirce, "Logic as Semiotic: The Theory of Signs," in: *Philosophical Writings of Peirce* (New York: Dover Publications, Inc., 1955), pp. 98–119.

we could do were we to disintegrate a painting into its component colors and then use them in the creation of a new color composition. The artist of this project is most fascinated by cognition, by the fact that biological material and DNA profiling are so open to manipulation that in fact there is no difference between this or any other medium, especially digital ones where we can extract components and re-compose them without damaging or irreversibly changing them during the manipulation, and by doing so we can create whole new entireties and repeat the procedure over and over again. The world becomes a computer game, and the body is no longer materiality or substance occupying space, weight, firmness, and transitoriness, *res extensa*, mortality; likewise, the body is not a unique subjectivity, but a file in a graphic program: infinite, manipulable, a photoshop drawing, optionally open, a hero with innumerable lives, a divisible and reversely constructed non-materiality, the opposite of entropy, complete reversibility, multiplicity, an avatar for anybody, for any identity (Vanouse, Simpson, whoever and whatever), infinitely changeable into something—a human or a worm, human-bacteria, pure evasion, evasiveness, escape, exodus of the owner from its own genetic profile—whose, what?

Shaking the indexical authority which ties the DNA profile as a sign tightly to its own object, to the human being who is being profiled, to whom this biological material belongs, the sign moves to the other pole, where the connection to the object is not essential; the object might not even exist, so what becomes important instead is the interpretation of a sign based on convention, therefore the primary thing here is *social codification*. At this pole the sign becomes a *symbol*. The symbolism in DNA profiling is of interest to Vanouse since he connects it to racially motivated stereotypes and prejudices. "Controversial criminal cases show that today's focus on genetic pool opens the door to racially motivated clichés and prejudgments that are mixed together with the suspects' genetic profile" is written in the foreword to the exhibition *Fingerprints*. If, on the one hand, the absolute authority of DNA profiling is established in a society in such a way that shows itself as true transparency through which we can see the truth, then we are presented with the regime of *perspicere par excellence*; on the other hand, if we believe Vanouse and Hauser, the exact opposite proves to be the case—the bond of the DNA sign with the suspect is questionable, thus the space opens for *proicere*, above all for the projections of social ideologies. The *Suspect Inversion Center* thus combines two scopic regimes, *perspicere* and *proicere*. As with machines for visibility (microscope, telescope and television), genetic inscription also helps us see what is invisible to the naked eye but nevertheless already here; we can understand it as piercing the surface, penetrating into inwardness, even to the utmost inner essence, to the real substance of the body. At the same time, genetic inscription is a typical case of the sort of projection which Zielinski sees in machines used for creating pictures (Ger. *Bild Maschinen*), like camera obscura, laterna magica, diorama/panorama, cinema.[7] The DNA visual display of the profile is namely a

---

[7]Zielinski, *Entwerfen und Entbergen*, p. 13.

form of transfer, translation, and projection onto another carrier, into another material. But the regime of projection does not simply mean the technical transfer of the picture, but also its active transfer, which is more than an intervention; it is projection in a manner similar to the way drama is constructed and or the way magic works. Vanouse does not want to be the cold observer who through the electrophoresis expedient in the genetic diagram sees an essence concealed even to the eye but uncovered here as the truth of the body. His role is active, his inter-vention constructive; instead of *un-covering* veils that obstruct the object of observation, he uses the technique of *creating*, not exactly as the creation from nothing, nor even in keeping with the principle of *proicere*, but as a process of breaking and decomposing and then composing and joining, of hybridization and even the alchemical technique of transubstantiation.

What restores the function of the *provability* of the DNA profile is the fact that it is a sign that substitutes its object (body), represents it, and stands for it as its *print*. This sign is supposed to prove the *presence* of something that is *absent* in the sign. That which is absent is thus supposed to become present. Re-presentativity, *re-peated* presentation means to show again, once more; the second time the pre-sentation must be *identical* with the first, the mark must be *genuine* and *credible*. Such a mark expands in a dimension of truth/non-truth that is superfluous (as acknowledged by Austin and the followers),[8] as the semiologic level (formal, expressive, occurring) which is tied to the semantic level (the level of meaning, content, sense) cannot possibly join with the outside-referential level (reality as thingness),[9] since between them there is a rupture, a crack, a *différence*. The essential predicate of a mark is iteration. Every mark is at the moment of its

---

[8]When he was thinking about utterances, the British philosopher of language John L. Austin (*How to Do Things with Words?*, 1955) discovered that the question about truth/non-truth, which had been an eternal question concerning language, is meaningless, since when I say: "It is raining outside" the question to ask is not: "Is it really raining outside?" but rather what is the force of my utterance and what have I achieved by it. For Austin the question thus shifts from constative statements to speech-acts. Austin supports the notion that reality is produced at the semiologic level, in the medium and with it; therefore we cannot distinguish between the "fictitious" and the "actual" when referring to outside reality. After Austin, John R. Searle was explicitly devoted to the question of truth/non-truth by opening up the question of the fictional discourse ("The Logical Status of Fictional Discourse," 1975). If fiction "pretends" to refer to some reality outside itself and so uses such references, so too in "realistic" discourses we don't see the reality that supposedly exists out there; we always have to deal instead just with the one that is in front of us. The examples of realistic and fictional discourse show that we actually always have in front of us just the reality of the discourse and not that of the outside-discourse reality, even though some dis-courses present themselves as credible in the relationship to the reality out there, as its proof, for which we have no guarantee whatsoever, except the sole one expressing the medium itself. Thus there is no difference between them—reality is always established at the level of the discourse.

[9]Even in the contemporary philosophy of society, authors are fascinated by discovering the dominance of *proicere* in places where *perspicere* was supposed to be operating. For instance, in 1996 Bourdieu said that "television, which claims to record reality, creates it instead." Pierre Bourdieu, *On Television* (New York: The New Press, 1998), p. 22.

construction separated from its source and also from its reception, therefore it can never be identical with what it is, i.e. itself.

We are discussing the question of identification, but we have not yet discussed the issue of identity. Martin Heidegger begins his lecture "The Principle of Identity" with the usual formulation of the principle of identity that reads "A = A," which is considered the highest principle of thought. He asks whether it is possible for this formula about equality to say that one is equal to another. "Obviously not," he replies. "That which is identical ... means 'the same.'"[10] If somebody repeats himself, he speaks in a tautology. The formula says: "every A is itself the same,"[11] or rather "every A is itself the same with itself."[12] Heidegger is aware that "[s]ameness implies the relation of 'with,' that is, a mediation, a connection, a synthesis: the unification into a unity."[13] This is how and why Western thought has comprehended identity as a unity. In this regard Heidegger reminds us that since speculative idealism it is no longer possible to represent the unity of identity as mere sameness and to disregard the mediation that prevails in such a unity. It was Hegel who first discussed "A = A" as a principle of contradiction; for him identity is dialectical. Derrida radicalized the notion of the rupture in the unity of identity and offered the concept of *différance*, speaking about the differential mark and the recognition that nothing can ever be identical to itself.

Derrida discusses the medium of writing and argues that a written sign is not a "progressive extenuation of presence" or supplementation (continuous modification) as it is in Condillac; instead, writing is determined by the break in presence, *différance*. A written sign breaks with its context, which includes the moment of inscription, the presence of the writer, the entire environment, and the writer's intention animating the inscription at a given moment. This holds true even for oral utterances, on which point Derrida opposed Austin, who did not doubt that the "source" (Derrida's term for Austin's "utterance-origin") of an oral utterance in the present indicative active is present to the utterance and its statement. There is a *différance*: "the irreducible absence of intention or attendance to the performative utterance, the most 'event-ridden' utterance there is, is what authorizes me ... to posit the general graphemic structure of every 'communication.'"[14] This force of rupture is tied to spacing [*espacement*] and iteration or repeatability, both of which constitute the written sign. The written sign possess a characteristic of being readable even if the moment of its production is irrevocably lost and even if the moment of reading is delayed. The very identity of a written sign (its iterability, its repeatability) does not permit a sign to ever be a unity that is identical to itself. The unity of signifying form only constitutes itself by virtue of iterability, the possibility

---

[10]Martin Heidegger, *Identity and Difference*, trans. Joan Stambaugh (Chicago: The University of Chicago Press, 2002), p. 23.

[11]Ibid., p. 24.

[12]Ibid., p. 25.

[13]Ibid.

[14]Jacques Derrida, *Limited Inc.* (Evanston, IL: Northwestern University Press, 1988), pp. 18–19.

of being repeated in the absence of its referent and the absence of a determinate signified or the intention of actual signification. The identity of signifying form is the "division or dissociation of itself."[15] For the same reason (the reason of iterability) Derrida opposed Austin's claim that a signature tethers the written utterance to its source. Derrida does not deny the occurrence of the absolute singularity of the signature as an event; this is the everyday effect of signature: "But the condition of possibility of those effects is simultaneously, once again, the condition of their impossibility, of the impossibility of their rigorous purity. In order to function, that is, to be readable, a signature must have a repeatable, iterable, imitable form," and this makes it the same kind of mark as the ones described above. It must be detached from the present and singular intention of its production: "It is sameness which, by corrupting its identity and its singularity, divides its seal."[16]

Perhaps it sounds bold to apply Derrida's theory of a differentiated mark to body prints, but it is often said that a fingerprint or a genetic fingerprint is a signature left at the scene of a crime. The written signature could be compared to other sorts of graphemes produced by bodies. There is not much sense in speaking about intention regarding a fingerprint, or at the very least there is no information available regarding intention. Regardless, if we agree with Derrida that the context of inscription is irrevocably lost in any mark, the issue of intention is irrelevant because, given the rupture with the source, there is not any difference between the identity of a written sign and a body print. Furthermore, body prints have a repeatable form; a fingerprint actually is quite similar to a written signature. It presents an absolute singularity, and at the same time it is repeatable. The signified is determined—the mark leads to a particular person, her individuality, her singularity. The mark is not only her signifier, but even her identifier. The meaning is shallow and deep at the same time—it only leads to her, nothing more, but nothing less, it is all she is. The signified is her identity. Still, it is a mark, torn from its context, the body, functioning on its own. As a mark it is subjected to differentiation, it is a *différance*, a "non-self-identity."[17]

Even if body prints are linked to the singular identities of the bodies that produce and therefore exhibit themselves as examples of pure self-identity, they are nevertheless torn from the context of their production and therefore are graphemes, which function as differentiated marks. There is one dimension which we would like to expose that Derrida does not discuss: body prints, like signatures or handwriting, are tracks of the body. Although a signature breaks with the presence of the inscriber, it also reveals the rhythm, the force, the dynamics, vehemence or modesty of the inscription; and even if all of this can perhaps be faked, handwriting is a trace of the body, a print of the extended ink-stained fingers on the surface. A finger- or a palm-print, even a print of a shoe, tells less about an individual's psychology or behavior (though some information can still be communicated), but it surely

---

[15]Ibid., p. 10.

[16]Ibid., p. 20.

[17]This is Derrida's description of *différance*. Ibid., p. 145.

comprises a physical link to the body, even if it is essentially ruptured, since the body is absent and the reading delayed. Additionally, our bodies not only leave printed traces in the spaces in which they appear, but they also leave smells or "smell traces," micro-organisms or "microbiological traces," and organic matter produced by the body. For some of these "physical traces"[18] left by bodies, we have developed technologies of visualization (while some are still used rather "primitively" in the sense that direct reading, i.e. recognition, is performed by other living creatures that already possess the ability to do so, like dogs or pigs), whereas body traces themselves have been ruptured, torn from their context, and the visualizations possibly produced are themselves as ruptured as the marks themselves. Genetic blueprints are thus especially interesting marks—they are doubly ruptured. Their identity as marks makes them subjectable to manipulation. This is what Vanouse has performed: he "misuses" their identity as marks and treats them as a grapheme, the signifying form of the sign. He has transformed the signifier (originating from his organic material or body trace) so as to make it into another signifier, one that is tethered to a particular individual and thus supposedly constitutes a unity that is identical with itself. However he does not transform the physical trace itself, the trace that "has been read by Vanouse." Notably, the O.J. Simpson case had issues with the purity of the traces themselves. If anybody were to take over the job of re-reading the trace analyzed by Vanouse by using the standard method of DNA blueprinting, she would come to a different conclusion, i.e. she would discover or establish the link to Vanouse's and not O.J. Simpson's body. Despite the incredible nature of Vanouse's methods, artists have demonstrated that even such credible marks as genetic fingerprints are only graphemes. Whereas iterability is present here not only in the repeatability of the signifying form, but also in production, traces could be reproduced and will in that case ensure the same results, unless the method ends up being "misused" as it is by Vanouse. What makes this particular example interesting is that eventually there is even a third level of iterability possible in certain cases: the very body traces that are required for genetic blueprints could in particular cases and under particular conditions simultaneously be body elements from which an entirely new, *genetically identical* body to the one traced and signified is *re-produced*. This iterability however makes little sense if we acknowledge the role that cultural *milieu* plays in the process of identification. In other words, the iterability of identity *par excellence* proves to be a misfire in achieving identity.

Mice constantly pee and thus leave a trace that enables the eagle to find them. A human is actually not much different. Not only does it secrete, but the body itself is constantly spread around through dead skin and other cells left behind. The body prints that we particularly focused on in this chapter are media depictions produced by body particles left in the environment. But in the environments that are

---

[18]The term "physical traces" seems a bit clumsy since even the prints are physical impressions, but they are rather impressions in another matter while with physical traces we mean particles of the body left in the location.

particularly rich with microorganisms (such as soil), these particles soon become useless as identifiers because the microorganisms literally digest them, so that parts of DNA get erased and biological material becomes non-identifiable. In this boundary zone, the body (not necessary a dead one, but a dispersed one) switches over into another process, that of decay. This is a process of transformation into energy and other forms of life. We participate in this process continuously. The body dissolves. Identity dissolves into non-identity. In such a manner, this non-identity happens to constitute identity itself, though at the same time the body constantly re-establishes its identity, recreates it, but not from scratch. This is a constant process of regeneration, the regeneration of identity and the regeneration of the body. Regeneration is to be recognized as constituting the essence of life.

Chapter 3
Solution of Life

Abstract Knowledge of life is the ground for defining life and death. These definitions ensure the political power over the body. With the development of technical devices that enable to see more and more and to "touch" the scales which the naked eye cannot see, the body is becoming an extensive milieu and knowledge of life changes, as we argue in the third chapter. Today, life can get dissolved, dispersed, diluted, or delayed. Precisely the ascertainment of the interrelations of the processes of living and dying affects the comprehension of body and this increases the biotechnological power of intervention.

Ontologically, life is to be understood as a biological process occurring in living creatures, in which life functions are being performed, such as birth, breathing, growth, nutrition, secretion, reproduction, and death. In accordance with this, deadness could be ascribed to an entity in which these processes do not occur. This might seem fine and non-problematic; in practice, however, especially in our biotechnological present, the criteria for life have become lax. For instance, as regards the human species, they are subject to legal regulations. Ascertaining the death of a human individual is an issue of medical diagnosis, originating from social consensus about existing definition of life and death. This is legally determined on the basis of a knowledge of life and its processes. The point is that no criteria could be understood as objective for or essential to the matter being observed; all criteria are subject to politics. Criteria originate from the comprehension of the optimal quality of the matter and that of the minimal conditions; they express the relationship to the matter and are linked to the possibility of its handling, even the power to render the optimal state, which is, last but not least, related to the support of techno-science. Furthermore, the politics connected to bodies that have been proclaimed dead exhibit not only power over dead (thanato-politics), but also over life and bodies.

In 1959 two neurophysiologists, P. Mollaret and M. Goulon, published a study in which they defined a novel, fourth form of coma (Fr. *coma dépassé*), which comprises not only relational life functions (in the classical coma, significant is the loss of such functions: consciousness, motion, sensibility, thermoregulation), but

P. Tratnik, *Conquest of Body*, SpringerBriefs in Philosophy,
DOI 10.1007/978-3-319-57324-3_3

also the complete abolition of vegetative life functions. In the case of this type of coma, life is prolonged by means of re-animation therapy until the heart is able to beat. The researchers conclude that the techno-scientific problem of re-animation has in this case lead to a necessary redefinition of death. Up to that moment, two traditional criteria defined death: the cessation of the heart and the cessation of breathing, criteria which became obsolete with the concept of *coma dépassé*. In 1968 a Harvard Medical School committee determined new standards that are today known under the term "brain death." As analyzed by Giorgio Agamben, the vague sphere of *coma dépassé* (floating in a wavering uncertainty between life and death) was overcome by defining irreversible coma as the new criterion of death.[1] The definition of brain death has since been legally recognized, and when diagnosed with the death of her brain, including the brainstem, a patient is considered dead, even if she continues to breathe thanks to techniques of re-animation (in the UK, irreversible brain stem dysfunction is accepted as the indicator of death). The legal definition of death changes over time according to the knowledge we have of life, and this knowledge is not the same worldwide. This demonstrates that clinical death does not mean the biological death of an organism as well. Agamben concludes: "life and death are not properly scientific concepts but rather political concepts, which as such acquire a political meaning precisely only through a decision."[2]

The issues of defining life and death have become increasingly relevant in the age of biotechnology because of various technological interventions into bodily and living processes, which have reopened questions as to what life is, what its essence is, how it is to be measured and determined, what consciousness is, how to treat patients with little or no consciousness, etc. There are no simple and definite answers to these questions, and doctors do not always agree. Decisions now have to be made about what do with living bodies that have socially died as individuals, first by society, then by doctors and families. Any living or dead body is subject to political power, which is not a sovereign power of the individual himself or herself. Especially in capitalist society, questions about power over the body are additionally related to questions about body ownership, which have gained new dimensions with the rise of biotechnology, in which a third party is able to patent the engineered construct for which somebody else's biological material has been used. With the commodification of body elements such as cells, a new chapter in the bio-market has begun.

Margaret Lock has investigated the politics and treatment of patients she calls "good-as-dead" because they are a useful source of material for organ transplantation and medical research, including brain dead patients, persistent vegetative state patients, and minimal consciousness patients. She notes: "In an era when commodification of the body is rampant—as source of intellectual property, financial gain, and therapeutic tools—certain individuals, notably the socially

---

[1]Giorgio Agamben, *Homo Sacer. Sovereign Power and Bare Life* (Stanford: Stanford University Press-Stanford, 1998), p. 162.

[2]Ibid., p. 164.

disenfranchised and the socially dead, are increasingly vulnerable to being counted as good-as-dead. They are judged as having little or no value—on the contrary, as being a drain on society—and therefore, it is often argued, better use could be made of their body parts in the utilitarian world of medical science."[3] Elsewhere she notes that in cases in which brain dead patients (in former times they were called living cadavers, because they were kept alive with the help of artificial respirators) do not become organ donors, the process of the preservation of *life* is very short, while on the contrary, in the case of donation it can be prolonged for quite some time. There resides here a novel, technologically supported death that occurs when brain dead patients die at the moment they are disconnected from the apparatus, and this death is in fact their second death.[4]

Some doctors claim that death is the cessation of the heart, but there are people who have survived a period of such death. In fact, surgeons stop the heart from beating during heart surgery—after a while in such a state, however, the patients would suffer brain damage. The criteria of death do not apply to suspended animation, but neither is it a completely mechanic operation. If it lasts too long, effects take place in the body that could become permanent. In other words, death gradually progresses throughout the body. It is not easy to determine *the* death of a person, since death dissolves into various sorts of deaths (social, biological) and is even constituted of several particular deaths (brain, heart, cells, consciousness) or partial deaths. It is actually life which dissolves into death gradually. We will pay more attention to this issue in a moment. Here we would rather proceed with the idea that a person and even an organism do not die at once. Terminology attests to some differences—if personality includes social life, an organism is a biological complex of life, for which we also claim that it passes through phases of dying. The bodies of brain dead patients continue to live. There are levels of autonomy: some unconscious bodies can breathe independently, some low conscious people can even chew on their own. The greater autonomy they have, though, the more difficult the decision to end their lives becomes and the more obviously such a decision ends up being tantamount to murder. If left just as they are, these bodies would die, but since we have the technological means of doing so, we have the option to keep them alive, thus the decision to preserve life stands in opposition to the decision to let death occur. Allowing someone to die has become a matter of power. Because life can be artificially prolonged, letting someone die means killing that person intentionally. But the goal of society is to make live, which is not just a senseless act—it is to "protect" society itself from extinction in the very effort to preserve the human species while consolidating, strengthening, and empowering a particular society. It is not strange that impatient research-engineering in "biology" goes hand in hand with finding solutions for how to travel throughout the universe and

---

[3]Margaret Lock, "On Making Up the Good-As-Dead in a Utilitarian World," in: Sarah Franklin and Margaret Lock (eds.), *Remaking Life & Death. Toward an Anthropology of the Biosciences* (Santa Fe: School of American Research Press, Oxford: James Currey Ltd, 2003), p. 191.

[4]Margaret Lock, "Twice Dead: Organ Transplants and the Reinvention of the Death," in: Miriam Fraser and Monica Greco (eds.), *The Body. A Reader* (Oxon, New York: Routledge, 2005), p. 263.

colonize other planets. It is all about the survival of human species—surmounting dangers that come from the universe, from global warming and from other societies.

If we take a perspective similar to the one we adopted in the first chapter, if we epistemologically "move" into the body and instead of remaining at the body scale observe the life of the body at the microbiological scale, then the gradual nature of the death of the body becomes readily apparent. Cells in the body do not die with the cessation of the heart. For us, who are so hung up on visuality, it is shocking to know that hair and nails still grow for a while. Biological processes within the organism gradually end. And today, when we become lost in reverie at the thought that we can re-develop a whole organism from a single cell, it is quite striking to realize how the complexity of life can be re-established after the death of a body. Recently, we have become so overweening because of the techno-knowledge power we've gained over the living that we have actually started to consider the possibilities of intervening into the process of dying in order to re-start life. One of the entrances to such thinking comes from our knowledge of the cell. In dealing with cells, the boundaries between life and death become even more ambiguous. This is not only the case because something continues to live even when it does not fulfill the crucial criterion for life that we apply to humans (that is to say, consciousness). At the level of cell, death is difficult to determine; the more we know, however, the easier it becomes to manipulate life.

Metabolic activity within the cell can be suspended, meaning the cell can be preserved alive even when it is not metabolically active. Skin or cartilage cells can be stored at cold temperatures (4 °C or 39.2 °F) at which they remain alive for a while, though they do not multiply, eat and secrete—as it were asleep. They are in a state of hibernation. After a month, a majority of these cells will still be alive, but hibernation has exhausted them, and they have become weaker. When transferred back to living conditions (37 °C or 98.6 °F), however, they re-activate their metabolic activities. Coldness also slows down the process of decay—the slowness of decay as the process of death is in this sense also a slowing down of the process of birth or re-birth, i.e. the process of life. Obviously, we are speaking about some sort of norm of life: life is considered as life in its "normal" state, while any other state is taken to be temporary, conditional, life at its minimum. Or the state of life in deficient conditions is not really reckoned as life, but rather as its suspense or delay. This supposed "normality" defines the essence of life and thus the criteria for it, which will in other cases determine its existence or non-existence, even its presence or absence. If we were to accept the definition that metabolic activity defines life, then in case of metabolic inactivity we would need to say that the cells are dead. This gets even more radical in the case of the preservation of the cells in liquid nitrogen at −130 °C or −202 °F. Organic matter is preserved unharmed, since this medium is cytoprotective—it prevents the formation of ice crystals, therefore this technique enables the *survival* of living entities like cells. All of the living processes of the organism cease. The state could be preserved infinitely. It is not a state of resting, but an actual cessation of life. Yet it is not death, because the cessation of life is not irreversible. After the "normal" state is re-established, the cell reactivates and again performs the biological processes. *Life literally survives death*. If life and death were considered to be two *different* states that mutually exclude each other, then this could not be possible. Therefore, life has this difference within itself, as does death.

The dialectical identity of life would consist in lacking deadness (the nothing of deadness), while deadness would be the nothing of life—life and deadness thus stand in opposition, they are a being and a nothing, but each contains difference within itself. Such would be the Hegelian comprehension at least. This takes us further: life needs to be diluted because absolute life would be the same as absolute deadness; it would have no difference in itself. This is how Hegel reflects upon light and darkness: "Pure light and pure darkness are two voids that amount to the same thing. Only in determinate light (and light is determined through darkness: in clouded light therefore), just as only in determinate darkness (and darkness is determined through light: in illuminated darkness therefore), can something be distinguished, since only clouded light and illuminated darkness have distinction in them and hence are determinate being, *existence*."[5]

Death is present in life and life in death. There are several perspectives we can take into consideration in this regard. It is this awareness which grounds the power to overcome the limitations of life.

Not long before Hegel's reflections Xavier Bichat, a French physiologist, commented upon life in relation to its opposite, death, and introduced a quantitative comprehension of life: "The definition of life is to be sought for in abstract considerations; it will be found, I believe, in this general perception: *life is the totality of those functions which resist death*. Such is in fact the mode of existence of living bodies, that every thing which surrounds them tends to their destruction."[6] According to Bichat, a permanent principle of reaction is the principle of life, such that the general phenomena of life are "that constant alternation of action on the part of external bodies, and of reaction on the part of the living body."[7] Bichat thus recognizes that in the infant the reaction is greater than the action, while in the adult an equilibrium is established, and in old age the reaction of the internal principle is diminished: "The measure of life then, in general, is the difference which exists between the effort of external powers, and of internal resistance. The excess of the former announces its weakness; the predominance of the latter is an indication of its strength."[8] In the twentieth century Georges Canguilhem, Michel Foucault's teacher, interpreted Bichat in these terms: "Bichat locates the distinctive characteristic of organisms in the instability of vital forces, in the irregularity of vital phenomena—in contrast to the uniformity of physical phenomena."[9] On the basis

---

[5]Georg Wilhelm Friedrich Hegel, *The Science of Logic*, trans. George di Giovanni (Cambridge: Cambridge University Press, 2010), p. 69.

[6]Xavier Bichat, *Physiological Researches upon Life and Death* (Philadelphia: Smith & Maxwell, 1809), p. 1.

[7]Ibid.

[8]Ibid., p. 2.

[9]Georges Canguilhem, *Knowledge of Life*, trans. Stefanos Geroulanos and Daniela Ginsburg (New York: Fordham University Press, 2008), p. 122.

of this understanding of life, he developed his famous distinction between the normal and the pathological: "to live, already for animals and even more so for man, is not merely to vegetate and conserve oneself. It is to confront risks and to triumph over them. Especially in man, health is precisely a certain latitude, a certain play in the norms of life and behavior."[10]

Canguilhem reflects on health and illness as qualitative oppositions; whereas illness is already inscribed in health, it defines it as a threat and a component of health: "We shall say that the healthy man does not become sick insofar as he is healthy. No healthy man becomes sick, for he is sick only insofar as his health abandons him and in this he is not healthy. The so-called healthy man thus is not healthy. His health is an equilibrium which he redeems on inceptive ruptures. The menace of disease is one of the components of health."[11] According to Hegel, Bichat and Canguilhem, life is defined by death since the threat of death is the identity of life, thus life has such a difference within itself.

Let us take a short excursus to see how life is protected in the body and how it trickles through and passes into death at the same time that there is great fertility of life and intertwinement with the rest of the world, such that the body spreads into its milieu and vice versa. The body actually is not a static complex of functions isolated from the world, not even in biological terms. In the interiority of the body, living tissues exist in a safe, wet and warm environment, embraced, relatively distant and protected from the external (cold and dry) world. Skin is typically understood as a cortex protecting the body (its essence), which is to be found in its interiority. In this sense it is the outer membrane of the body, the main purpose of which is to protect interiority—that is to say, the living conditions assured there— from the intrusion of foreign species from the external world. The world is thus split into two parts: interiority (the body essence, its identity) and the rest of the world (the other). Such a comprehension is too simplified and for this reason misleading. The external layer of skin is a thin layer of epidermis, which is itself composed of three sub-layers.[12] The horny layer (stratum corneum) contains dead keratinocytes (the keratin, a protein formed from dead cells, protects the skin from harmful substances). Deeper there is a layer of living keratinocytes (squamous cells), and the internal layer of epidermis is where the basal cells continually divide, thus forming the keratinocytes. This process of stratification takes place within the body: cells formed in the interior of the skin travel to the external stratum, the last layer of bodily defense. Contrary to the tissues composed of living cells that exist in the interiority of the body, this layer is composed of dead cells. We could say that life traverses into death when making its way from the interior of the body to the colder

---

[10]Ibid., p. 132.

[11]Georges Canguilhem, *The Normal and the Pathological*, trans. Carolyn R. Fawcett (Dordrecht: D. Reidel Publishing Company, 1978), p. 179.

[12]Deeper in the body, skin is composed of thicker layers: dermis as the middle layer and subcutis as the deepest layer. Dermis also contains the pain and touch receptors and is held together by collagen (a protein made by fibroblasts—these are the cells that give the skin its strength and resilience). Subcutis protects the other organs from injury by acting as a shock absorber of sorts.

and drier milieu that encompasses it. In this sense, it is a boundary, life's limit. When cultivating these cells in vitro, this boundary is substituted by several artificial layers: first the limit of the scaffold (its encounter with air), then the petri-dish and finally the boundary of the incubator. In the body, it is fascinating to acknowledge that this limit of life (the layer of death) is nevertheless constituent of the very form of life. Moreover, one cannot comprehend the horny layer simply as an area of death. For life is condensed here. There are thousands of organisms that live here constantly. This is not a sphere where the body encounters the dry, unpleasant and non-advantageous world. It is rather a sphere of juncture. Furthermore, the micro-organisms that inhabit this area are not intruders; instead they literally compose the human body and constitute its identity. They can be even taken as one of the body prints we discussed in the previous chapter.

The genetic profile of an individual is actually a much more complex assemblage than a profile of just one species. But this is not a restricted identity. Can one still argue today for the idea of an intimate, individual body, an identity that is preponderantly distinguished and separated from the other world, especially if a human is to be acknowledged as being immersed in the world, where things reciprocally belong to each other in such a manner that they constitute the same flesh of the world? These microorganisms do not only live as components of the body, but also traverse from it into the environment, and vice versa; an exchange of living substances is constantly taking place. The outer layer of the body is thus something of a vibrant exchange market; it is a region of the dissolution of identity: interiority extends up to exteriority and exteriority enters down to interiority, although concepts of interiority and exteriority become senseless here. The body and the environment are intertwined, but even this is difficult to insist upon since in saying as much we still consider them as distinguished entities. The body disperses into milieu, and milieu spreads onto and into the body. Only recently have researchers started to seriously consider the complex interrelations between the human organism and microorganisms or the significance of these microorganisms for the world.

With this deconstruction of the notions of the boundaries of the body and the body as a closed distinguished substance, the idea of the wholeness of the organism loses a great deal of its sense. Not only is the body to be considered as a multitude, but also its notion should denote openness, exchangeability, its traversabilty and permeability. Its structure is, in short, rhizomatic. The body is a *rhizome*.

In discussing the dissolution of the body into the world and the issue of identity and otherness, we can again draw attention to Hegel and his consideration of something and an other: "The other moment is equally an existent, but determined as the negative of something—an *other*."[13] For Hegel: (1) something and an other both exist as something; (2) each is equally an other in its relation to its otherness and is preserved in its inexistence; and (3) both moments are determinations of one and the same thing—i.e. something. "The other which is such for itself is the other

---

[13]Hegel, *The Science of Logic*, p. 90.

within it, hence the other of itself and so the other of the other—therefore, the absolutely unequal in itself, that which negates itself, *alters* itself."[14] As the other of the body, the milieu is its dialectical negative, it is determined with the body, while if the body is to be considered as the other of the milieu, it is its negative, determined along with the milieu. The difference between the body and milieu is sublated (*aufgehoben*): they are both differentiated in themselves, they are determinations of one and the same. Furthermore, if the body and milieu are considered as something and other, there must be a limit to that something (either the body or the milieu), there must be a boundary between them. This comprehension is adopted by the discourse of immunology, to which we will return in the next chapter. A body is a temporary locus of a particular organization of life. It is not a closed entity, but rather a drizzling identity. The delimitation of the body is a construct, and it is therefore arbitrary where it ends up being drawn. The world into which the body is immersed and which it traverses is also the world into which life dissolves.

Because life and death also interrelate in a temporal relationship in such a way that the life of an organism is increasingly permeated with death until finally the body enters the process of decay, death is perceived as a limit of life. However, death as the end of life (as life's limit) has also been sublated. This is a situation already discussed by Hegel: "The limit which arises in this beyond is therefore only one that again sublates itself and sends itself to a further limit, *and so on to infinity*."[15] Hegel calls this kind of infinity bad infinity: "The bad infinity, especially in the form of the *quantitative progress to infinity*—this uninterrupted flitting over limits which it is powerless to sublate, and the perpetual falling back into them".[16]

This brings us back to the issue of the delimitation of life, which we have claimed is also constructed and determined by a governing discourse. Furthermore, today the power over life and death has become vigorous. Life can be turned on and off. With biotechnology the difference between life and death is not only sublated, but also mastered.

The possibility of manipulating life has become literal. The preservation of living material in liquid nitrogen is now a regular social practice in the case of bio-banks for umbilical cords. In societies where this is possible, when giving birth almost any mother can decide to preserve stem cells for future medical treatment. But the ambitions to manipulate life are even greater. Researchers find a challenge in turning the process of dying into a process of living. This would not be possible, however, if both were not considered determinations of one and the same thing.

One of the interests of biology today is in the process of apoptosis in multicellular organisms, so-called programmed cell death (in contrast to necrosis, a form of traumatic cell death). For researchers it is a challenge to observe the cell in life and as it dies, since all cells are primed for self-destruction and medicine wants to

---

[14]Ibid., p. 92.

[15]Ibid., pp. 189–190.

[16]Ibid., pp. 192–193.

make them die on cue: "The reordering of disease as a consequence of 'too much' or 'too little' of such death [cell death] and the conceptualization of this death as a mechanism with a spatially located apparatus and a temporally situated cascade suggest the possibility of these places and times as points of entry for therapeutic interventions."[17] When cells are in unfavorable conditions (such as those depriving them of oxygen), mitochondria responsible for the aerobic respiration of the cell are exploited by apoptotic pathways and instead of supplying cellular energy trigger a death signal, thus causing a chain reaction of cell death. In other words, mitochondria are responsible for diverse processes in the cell, from energy supply to cellular differentiation, signaling, cell cycling and growth, as well as death. Researchers now face the challenge of reprogramming mitochondria to produce energy instead of producing death.

The very recent acknowledgments of biological research have furthermore confirmed this essential linkage between life and death. Researchers acknowledge the double nature of stem cells in the organism, which confirms that the same grounds exist for regeneration and illness formation. They have begun to acknowledge that the very process which generates life destroys it as well. Within the living system, the vital force turns into a destructive force under certain conditions. Once we know what these conditions are, and once we gain an understanding of the processes taking place at a cellular level, we will be better able not only to comprehend life, but also to intervene into it by means of stem cell and genetic biotechnology, which today play a crucial role for biopower.

The current stem cell and cancer research conducted by Tamara Lah Turnšek and her team at the National Institute of Biology in Ljubljana acknowledges the double nature of stem cells in cancer. It demonstrates that the processes of regeneration fundamentally supported by stem cells are linked to processes of disease formation that start within the organism and depend on the inherent quality (age and viability and proliferation potential) of the cells, which is genetically programmed in each cell within a body. Additionally, it is dependent on the cellular microenvironment, as well as on external stress and environmental effects. Carcinogenesis has a dual role: on the one hand it generates new stem cell ability to develop in an organ, with lack of growth control and organization that differ from normal stem cells. On the other hand, tumor growth triggers the intrinsic systemic organism's response in recruiting normal, adult tissue stem cells aimed at tissue regeneration for resisting and curing the disease—cancer in this case. Tumors can indeed be looked upon as a never healing wound, due to the fact that tumor cells attract and recruit a number of immune cells and stromal cells that are supposed to have anti—tumor activity. However, the battle is soon lost due to the reprogramming of these normal cells within the tumor, which turns them into cancer-promoting cells facilitate tumor progression. Contemporary biomedical stem cell technology is trying to better understand the interactions between tumor and normal stem cells that would enable

---

[17]Hannah Landecker, "On Beginning and Ending with Apoptosis," in: *Remaking Life & Death*, p. 53.

intervention and eradiation of cancer and cancer stem cells in particular for their selective destruction, thus enhancing the power of normal tissue stem cells heal the diseased tissues. At present researchers are working on an approach to enhance the potential of normal mesenchymal stem cells with methods of genetic engineering that induce cell death in cancer stem cells. By finding the key genes, it would be possible to reprogram the vital regenerative forces by endogenous mesenchymal stem cells to fight this disease. Such are the regenerative body secrets that contemporary biotechnology is promising to reveal.

Chapter 4
Regenerative Body: Biopower with Biotechnology

Abstract In the last chapter, we discuss the paradigm of the regenerative body, which is particularly interesting since it has discovered the quality that enables us not only to distinguish life from mechanics, but also to intervene into life processes in order to "improve" or "rescue" the body from dying or aging. This is ensured by the quality of regeneration. Regenerative body generates an ultimate dream of the conquest of body: an immortal active life of a body in constant process of vitalization, with which the process of mortification is defeated once and for all.

In his accompanying essay to the translation of Roberto Esposito's book *Bíos*, Timothy Campbell asks: "What does the opening to *bíos* as a political category that humanity shares tells us about the other development that so decidedly marks the current biopolitical moment, namely, biotechnology?" He continues by noting that "missing is precisely a reflection on the role biotechnology plays for contemporary biopolitics."[1] In this chapter we draw attention to the growing importance of biotechnology for the power over life and body, and we analyze biotechnology as a contemporary biopolitical strategy. We comprehend biotechnology as a political technology which is investing in the body, improving its qualities, prolonging youth, taking care of health and reproduction. In such a sense, it preserves or protects life by helping to improve health, enriching the quality of life and enabling active ageing. It intensifies techniques of biopolitics and anatomo-politics (detected by Foucault) and implicates specially derived politics, engineering-politics and regenerative-politics, which demonstrate that there is power over life and the body in contemporaneity far exceeding the extensions and technological possibilities of power deriving from biological modernity. The possibilities for technological interventions into the "natural" are growing, and the difference between the natural and the technological is getting increasingly blurred. This marks the beginning of a new chapter in biopower, one that no longer belongs to biological modernity. The importance of biotechnology for biopower has been recently acknowledged in the lively debate over biopolitics: "The patenting of the human genome and the

---

[1]Timothy Campbell, "*Bíos*, Immunity, Life," in: Roberto Esposito, *Bíos. Biopolitics and Philosophy* (London, Minneapolis: University of Minnesota Press, 2008), p. xxxiii.

© The Author(s) 2017
P. Tratnik, *Conquest of Body*, SpringerBriefs in Philosophy,
DOI 10.1007/978-3-319-57324-3_4

development of artificial intelligence; biotechnology and the harnessing of life's forces for work, trace a new cartography of biopower. These strategies put in question the forms of life itself."[2] Within the growing interest in biopolitical issues related to the development of the life sciences, genomics has drawn quite a lot of attention, while on the contrary, the implications of the younger field of regenerative medicine has not yet been comprehensively discussed. This chapter is focused particularly on regenerative medicine as the knowledge-technology power that opens a new horizon for biopower.

Michel Foucault recognized an important historical shift regarding the relations between politics and life, which he located between the ancient era of sovereign power and the modern era of *biopower*, "when the life of the species is wagered on its own political strategies."[3] This moment is marked by a shift in power relations: "For millennia, man remained what he was for Aristotle: a living animal with the additional capacity for a political existence; modern man is an animal whose politics places his existence as a living being in question."[4] In order to consider the semantics of the term biopolitics and the relation between life as a "natural" issue and politics, one needs to refer to the ancient Greek (in particular the Aristotelian) lexicon to find the etymological origin in the Gr. term *bíos* (βίος). But the ancient Greeks used two terms to denote life: "*zoē*, which expressed the simple fact of living common to all living beings (animals, men, or gods), and *bios*, which indicated the form or way of living proper to an individual or a group."[5] The present tense *zo* means "I am alive, I exist" and the past tense (usually the case with the second past tense) *ebion* (meaning "I lived my life in a specific way") is an ancient form, from which came into existence the later present tense *bioo*: "The past tense 'EBION' and the derivative noun 'BIOS' were constructed in order to indicate a new notion about life, a notion more concrete and specific: i.e., the constant purposive and therefore complete, unchangeable way of life, to live a life, as Aristotle says, in a concrete mental way ... 'BIOS is a moral action.'"[6] Zoe generally refers to the existence of a living being, and *bíos* denotes qualified life. *Bios* is a duration of *zoe* and means rational life, thus it cannot be ascribed to animals.[7] In the classical world of the ancient Greeks, simple natural life is excluded from *pólis* ("to speak of a *zoē politikē* of the citizens of Athens would have made no sense").[8] However, Roberto Esposito relativizes the distinction between the two Greek terms denoting

---

[2]Maurizio Lazzarato, "From Biopower to Biopolitics," in: *The Warwick Journal of Philosophy* Volume 13, p. 1, http://cms.gold.ac.uk/media/lazzarato_biopolitics.pdf, 06-27-2011.

[3]Michel Foucault, *The History of Sexuality. Volume I: An Introduction* (New York: Pantheon Books, 1978), p. 143.

[4]Ibid.

[5]Agamben, *Homo Sacer*, p. 9.

[6]Michael Bakaoukas, "The Good Life. An Ancient Greek Perspective" http://ancienthistory.about.com/library/bl/uc_bakaoukas4a.htm, 12-29-2009.

[7]Ibid.

[8]Agamben, *Homo Sacer*, p. 9.

life because "every life is a form of life and every form refers to life."[9] He notices an interesting oscillation in the semantics of the Greek lexicon, namely "biopolitics refers, if anything, to the dimension of *zōē*, which is to say to life in its simple biological capacity, more than it does to *bíos*, understood as 'qualified life' or 'form of life,' or at least to the line of conjugation along which *bíos* is exposed to *zōē*, naturalizing *bíos* as well."[10] Furthermore he problematizes the concept of *zōē* and adds another term to the dualism of *bíos* and *zōē*: "*Zōē* itself can only be defined problematically: what, assuming it is even conceivable, is an absolutely natural life? It's even more the case today, when the human body appears to be increasingly challenged and also literally traversed by technology. Politics penetrates directly in life and life becomes other from itself. Thus, if a natural life doesn't exist that isn't at the same time technological as well; if the relation between *bíos* and *zōē* needs by now (or has always needed) to include in it a third correlated term, *technē*—then how do we hypothesize an exclusive relation between politics and life?"[11]

In the middle of the 1970s Michel Foucault, starting with the ancient Greek comprehension of life and its inclusion of "natural" life in political mechanisms (and with previous theories of biopolitics as well), re-proposed and redefined the concept of *biopolitics* in a much more complex sense than this had been done before.[12] Foucault outlined the difference between biopolitics as a politics in the

---

[9]Esposito, *Bíos*, p. 194.

[10]Ibid., p. 14.

[11]Ibid., p. 15.

[12]In a brief genealogy, Roberto Esposito sheds a great deal of light on the concept of biopolitics before and after Foucault (Esposito, *Bíos*, pp. 13–24) and traces the first wave of early discussions of biopolitics to the beginning of the 20th century in Swedish (Rudolph Kjellén 1905, 1916, 1920), German (Baron Jakob von Uexküll 1920), and English (Morley Roberts 1938) thought, in which it mostly referred to geopolitics and used an organistic, anthropological, and naturalistic approach, where the "naturalization of politics" took place in an analogical understanding of the state with its tissues as an organic whole (Kjellén and von Uexküll) and where the comparison between the defensive apparatus of the state and the immune system was discussed (Roberts). These early approaches show that "a politics constructed directly on *bíos* always risks violently subjecting *bíos* to politics" (Ibid., p. 19). The second wave of interest (appearing in France in the 1960s) demonstrates the modifications resulting from the epochal defeat of Nazi biocracy and the necessity of a semantic reformulation that is neohumanistic, although it finally resulted in weakening the specificity of the category, making it something like an "onto-politics." The third wave has taken place in the Anglo-Saxon world—and this is the one that is still ongoing. It emerged in the 1960s and was formally introduced in 1973 by the International Political Science Association, which opened a research site on biology and politics, and it was further marked by the foundation of the Association for Politics and the Life Sciences in 1983. This approach has a naturalistic character—its symptomatic value resides in the direct and insistent reference made to the sphere of nature as a privileged parameter of political determination. Esposito notices a considerable categorical shift with respect to the principal line of modern political philosophy: "While political philosophy presupposes nature as the problem to resolve (or the obstacle to overcome) through the constitution of the political order, American biopolitics sees in nature its same condition of existence: not only the genetic origin and the first material, but also the sole controlling reference. Politics is anything but able to dominate nature or 'conform' to its ends and so itself emerges 'informed' in such a way that it leaves no space for other constructive possibilities" (Ibid., p. 22).

name of life (politics of life) and *biopower* as subjecting life to the command of politics (politics over life). *Biopower* designates "what brought life and its mechanisms into the realm of explicit calculations and made knowledge-power an agent of transformation of human life."[13] In other words, biopower means "a number of phenomena that seem to me to be quite significant, namely, the set of mechanisms through which the basic biological features of the human species became the object of a political strategy, of a general strategy of power, or, in other words, how, starting from the eighteenth century, modern Western societies took on board the fundamental biological fact that human beings are a species."[14] Foucault recognizes the beginning of the age of biopower taking place with the end of the sovereign power for which "[t]he sovereign exercised his right of life only by exercising his right to kill, or by refraining from killing; he evidenced his power over life only through the death he was capable of requiring. The right which was formulated as the 'power of life and death' was in reality the right to *take* life or *let* live."[15] In the modern era of biopower, the social body has the right to ensure, maintain, or develop its life: "the ancient right to *take* life or *let* live was replaced by a power to *foster* life or *disallow* it to the point of death."[16] If before the sovereign was the one who was defended, now wars are waged on behalf of the existence of everyone, thus society is what must be defended; entire populations are mobilized for the purpose of wholesale slaughter, with underlying tactics of battle: one has to be capable of killing in order to go on living. Thus, the dream of modern power is genocide. Power is situated and exercised at the level of life, the species, the race, and the large-scale phenomena of population: "But this formidable power of death [demonstrated by the bloody wars since the nineteenth century] … now presents itself as the counterpart of a power that exerts a positive influence on life, that endeavors to administer, optimize, and multiply it, subjecting it to precise controls and comprehensive regulations."[17]

According to Foucault, this power *over* life was one of the basic phenomena of the nineteenth century in Western society, and it evolved two basic forms, which constituted two poles of development that were linked together. The first emerged in the seventeenth and eighteenth century and is a disciplinary technology—Foucault calls it the *anatomo-politics of the human body*. It "centered on the body as a machine: its disciplining, the optimization of its capabilities, the extortion of its forces, the parallel increase of its usefulness and its docility, its integration into systems of efficient and economic controls."[18] The other pole emerged in the middle or second half of the eighteenth century and "focused on the species body,

---

[13]Foucault, *The History of Sexuality. Volume I*, p. 143.

[14]Michel Foucault, *"Security, Territory, Population". Lectures at the Collège de France, 1977–78* (11 January 1978) (New York: Picador, 2009), p. 1.

[15]Foucault, *The History of Sexuality. Volume I*, p. 136.

[16]Ibid., p. 138.

[17]Ibid., p. 137.

[18]Ibid., p. 139.

the body imbued with the mechanics of life and serving as the basis of the bio-logical processes: propagation, births and mortality, the level of health, life expectancy and longevity, with all the conditions that can cause these to vary."[19] The supervision of these was effected through an entire series of interventions and *regulatory controls* that amounted to the *biopolitics of the population*. These two technologies were directed toward the performances of the body and with an attention to the processes of life, with the result that the highest function of this power over life "was perhaps no longer to kill, but to invest life through and through."[20] Both politics, (anatomo-politics and biopolitics) were, according to Foucault, techniques of power established in the course (anatomo-politics) and at the end (biopolitics) of the eighteenth century, and they were present at every level of the social body and utilized by very diverse institutions, such as the family, army, schools, police, individual medicine and the administration of collective bodies. There was a big difference between the era following the French Revolution and antiquity, namely in the fact that "death was ceasing to torment life so directly. But at the same time, the development of the different fields of knowledge concerned with life in general, the improvement of agricultural techniques, and the observa-tions and measures relative to man's life and survival contributed to this relaxation: a relative control over life averted some of the imminent risks of death."[21] The new regime supported the affirmative politics of life and over life: "Power would no longer be dealing simply with legal subjects over whom the ultimate domination was death, but with living being, and the mastery it would be able to exercise over them would have to be applied at the level of life itself; it was the taking charge of life, more than the threat of death, that gave power its access even to the body."[22] The political technologies that ensued—investing in the body, health, modes of subsistence and habitation, living conditions, and the whole space of existence—only proliferated.

Foucault discusses the issues of biopolitics in several of his lectures and papers. It is interesting that his first use of the term appeared in a 1974 lecture in which he emphasized the importance of biopolitics and recognized medicine as a biopolitical strategy: "for capitalist society it is the biopolitical that is important before everything else; the biological, the somatic, the corporeal. The body is a biopolitical reality; medicine is a biopolitical strategy."[23] The role of medicine and clinics is of great importance to Foucault's discussion on biopower and biopolitics. He con-ducted comprehensive research into the birth of the clinic from the middle of the eighteenth to the middle of the nineteenth century. With the coming of Enlightenment, death was subjected to the clear light of reason and became an object and source of knowledge for the philosophical mind. With the introduction

---

[19]Ibid.

[20]Ibid.

[21]Ibid., p. 142.

[22]Ibid., p. 142–143.

[23]Quoted in: Esposito, *Bíos*, p. 27.

of dissection rooms into the clinic in the middle of the eighteenth century, a new period began for medicine, which turned to the study of physiological phenomena. But there is a paradox in reading symptoms from anatomical perception: "A clinic of symptoms seeks the living body of the disease; anatomy provides it only with the corpse."[24] Thanks to the organization of the clinic in the eighteenth century, pathological anatomy (the technique of corpse observation) was allowed to open up a corpse right after the occurrence of death. This meant the latency period between death and autopsy was reduced such that the stage of pathological time and the first stage of cadaveric time almost coincided. The effects of organic decomposition were therefore virtually suppressed thusly: "Death is now no more than the vertical, absolutely thin line that joins, in dividing them, the series of symptoms and the series of lesions."[25]

In the late eighteenth century, Xavier Bichat introduced a new paradigm into medical thought, which replaced the former nosology based upon the principle of localization (understanding the illness of the body on the basis of organic prox-imity) with the principle of isomorphism in tissues, based upon the similarity and external adaptation of tissues, life characteristics and functions. Bichat imposed a diagonal reading of the body carried out according to expanses of anatomical resemblances that "traverse the organs, envelop them, divide them, compose and decompose them, analyse them, and, at the same time, *bind them together*."[26] He also recognized that when the pathological state is prolonged, the first tissues to be affected are those in which nutrition is most active (the mucous membranes), then the effects expand to the parenchyma of the organs and finally they reach the tendons and aponeuroses. Bichat ascertained that a disease is actually a process which "announces the coming of death."[27] Disease as the "proximity of death" is a process that indicates another process, one that is evolutionary: "the associated, but different process of 'mortification.'"[28] With this acknowledgment, death is no longer an instantaneous event, but rather a process. What Bichat actually acknowledged is "the permeability of life by death."[29] Foucault locates a shift in the comprehension of life related to the body with Bichat's contribution to pathologic anatomy. Particularly significant was Bichat's investigation of the body as a complex of tissues and his understanding that the analysis of disease can be carried out only from the point of view of death, "of the death which life, by definition, resists,"[30] whereas "[t]he morbid is the *rarefied* form of life, exhausted, working itself into the void of death."[31] For Bichat "[d]eath is therefore multiple, and

---

[24]Foucault, *The Birth of the Clinic*, p. 135.

[25]Ibid., p. 141.

[26]Ibid., p. 129.

[27]Ibid., p. 141.

[28]Ibid.

[29]Ibid., p. 142.

[30]Ibid., p. 144.

[31]Ibid., p. 171.

dispersed in time: it is not that absolute, privileged point at which time stops and moves back; like disease itself, it has a teeming presence that analysis may divide into time and space; gradually, here and there, each of the knots breaks, until organic life ceases, at least in its major forms, since long after the death of the individual, minuscule, partial deaths continue to dissociate the islets of life that still subsist."[32] Vitalism appeared against the background of "mortalism." Bichat relativized the concept of death, volatilized it, distributed it throughout life in the form of separate, partial, progressive deaths, deaths that occur so slowly that they extend even beyond death itself, but "from this fact he formed an essential structure of medical thought and perception: that to which life is *opposed* and to which it is *exposed*; that in relation to which it is living *opposition*."[33] Foucault is convinced that the irreducibility of the living to the mechanical or chemical is of secondary importance in comparison to this fundamental link between life and death.

This shift in the comprehension of death and life in biological modernity, however, was not accidental. Foucault acknowledges that epidemics were not the issue at the end of the eighteenth century anymore, but rather "endemics, or in other words, the form, nature, extension, duration and intensity of the illnesses prevalent in a population."[34] These were the illnesses that were difficult to eradicate and that had become permanent factors which sapped the population's strength, shortened the working week, wasted energy, and cost money (in the sense that they led to a fall in production and their treatment was expensive)—thus these were the phenomena affecting a given population. Therefore, death was no longer something that suddenly swooped down on life as in an epidemic, but instead became something permanent, something that slips into life, perpetually gnaws at it, diminishes it and weakens it.[35] This problem is a biopolitical one, and it became an important issue at the time of industrialization (in the early nineteenth century), particularly in the form of the problem of aging, when individuals fall out of the field of activity. Herein lies the significance of medicine for biopower, since biopower "is continuous, scientific, and it is the power to make live."[36] The "power is decreasingly the power to take life, and increasingly the right to intervene to make live."[37]

Intervention for the aim "to make live" has gained tremendous extensions with the advent of biotechnology in the last half of the twentieth century, which focuses on *technological* intervention and thus the *"artificiality"* of life as well. Since medicine has gained the power to intervene and engineer, living organisms can no longer be perceived as self-contained and delimited "natural" bodies but rather as

---

[32]Ibid., p. 142.

[33]Ibid., pp. 143–144.

[34]Michel Foucault, *"Society Must Be Defended." Lectures at the Collège de France, 1975–76* (17 March 1976) (New York: Picador, 2003), p. 243.

[35]Ibid., p. 244.

[36]Ibid., p. 247.

[37]Ibid., p. 248.

constructs composed of heterogeneous and exchangeable elements (e.g., organs, tissues, DNA). The involvement of technologic manipulation of the body in medicine has only been increasing since the middle of the twentieth century, and today biotechnology has become a significant supporting technology for medicine. The questions concerning the "natural foundations" of life and how these can be distinguished from "artificial" forms of life have become topical because of bio-scientific discoveries and technological innovations.[38] The ancient relation between "natural" life and politics that is an issue of the contemporary philosophy of biopolitics has become complex if one considers how the political encompasses sets of problems that were once understood as natural and self-evident facts but which are now open to technological or scientific intervention. Within "technocratic biopolitics," as Thomas Lemke calls them, the growing significance of genetic and reproductive technologies have raised concerns about the regulation and control of scientific progress. The results of biological and medical research and their practical applications have demonstrated how contingent and fragile the boundary between nature and culture is, but this has only intensified political and legal efforts to reestablish that boundary. It has thus become necessary to regulate which proce-dures are acceptable and under what conditions they may be deemed so.[39]

At the turn of the millennium, biotechnology gained a new promising branch: tissue engineering. The turn away from the computer paradigm of life that underlies genetics must have had to do with disappointments regarding the results of the Human Genome Project, which informed us that humankind has only 30,000–40,000 genes (and not 120,000 or 140,000 as was predicted) and that the biological difference between humankind and other species is relatively small, as is the genetic difference between individuals. The quantities (of differences and elements) must have different qualities than we had thought of before, meaning that a very small number of genes determines a very wide palette of features. These results did not tell us very much, and that was precisely the problem. How much is relatively small, what do we really know about them, what are our capacities to intervene with a given knowledge, etc.? These questions have only started to arise. So it seemed as if we were very far from comprehending how this system works and how to master it, since we have thus far recognized not much beyond the scale of the genetic determination of identity. At the turn of the millennium, the high listed shares of the biotechnological companies steeply fell on the stock exchange. The disappoint-ments that succeeded such great expectations demonstrated that the ultimate tele-ological aim to which humankind had been ascending had proven to be an empty point. As we have experienced in every other historic situation, with every revo-lution or war that promised freedom, this extremely problematic notion of *telos* is slippery. The initial impulse and great aim of this progression, together with our sense of it as a progression or whole constructed totality, collapsed instantly.

---

[38]Thomas Lemke, *Biopolitics. An Advanced Introduction* (New York, London: New York University Press, 2011), p. 27.
[39]Ibid., p. 26.

However, the era of biotechnology has not ended. There are other branches which promise great solutions today, and there is the knowledge-technology of genomics, which is jointly made up of novel body explorations and engineering. Nevertheless, the paradigm of the regenerative body discussed here has some distinctions in comparison to the that of genomics, which was significantly marked by the digital age and computer culture that was consolidated during the second half of the twentieth century. Tissue engineering is not based on the idea of programming, but is rather constructional: it is a sculptural and architectural branch. Tissue engineering introduces an alteration of the perception of life and body. It replaces the computer notion of a software life program and hardware body materialization with the notion of a body as a construction site, in which parts are installed on the skeleton.[40] But the very novel cognition it delivers is that these parts can be developed or just regenerated from and by the body itself. The elements that will form the parts will be taken from the body, improved, and then returned to the body in a better form. This has several important implications, including the probability that we are not going to need organ donors and that we will have chances to preserve our own bodies and improve them, enhance their immanent capacities and make them live longer and in a healthier way. The idea that the effects of intervention into our very genetic origin would only take place after the birth of a new organism is simply obsolete in comparison with current options promising the improvement of what is already here, of the situation in which we already find ourselves, i.e. the life we now live. This is precisely what is promised by the function of regeneration.

Tissue engineering manipulates the body's own cells in order to compose or regenerate tissues or body parts. Recently, the big promise has been related to stem cells. Tissue engineering, particularly stem cell engineering, has presented a new hope in the last decade, albeit one that is more realistic and less utopian in comparison to the promises of genomics.[41] Primarily, tissue engineering emerged as a response to transplantation problems, mainly to the response of the immune system, specifically the rejection of allogenic tissue.[42] Tissue engineering options (first the manipulation of skin or cartilage cells) became the center of biotechnology around

---

[40]However, it is quite a generalization to distinguish between the two so radically. It is worth mentioning that today tissue engineering and genomics have joined forces in efforts like the attempt to reprogram skin cells to become equivalent to embryo stem cells, and thus they do not represent two different politics as such.

[41]Herbert Gottweis reviews the reception of the life sciences in the last decades of the twentieth century as follows: (1) the 70s present the phase of hopes and fears, (2) the 80s the phase of exaggerations and (3) the 90s the overtaking of fantasies by contradictory realities. See Herbert Gottweis, "Genetic Engineering, Scientific-Industrial Revolution and Democratic Imagination," in: Gerfried Stocker and Christine Shöpf (eds.) *Ars Electronica 99. Life Sciences*, (Linz: Ars Electornica, Festival for Art, Technology and Society, 1999), pp. 122–134.

[42]In 1967 the first heart transplant was performed on Louis Washkansky in Cape Town, South Africa by cardiac surgeon Christiaan Bernard. Although Washkansky died eighteen days later of double pneumonia, the surgery was considered a success because the heart was beating on its own, without the assistance of a machine.

2000, with one of the breakthroughs occurring among biotechnologists at MIT who grew an ear with human cells in the body of a laboratory mouse (the results were published in 1995). The prototype human ear made of polyester fibers and human cartilage cells was transplanted onto the back of the mouse. The mouse tissue itself cultivated the ear during the growth of the cartilage and it finally completely replaced the artificial fibers. The technology immediately promised that at some point in the near future we would routinely be able to re-grow ears, noses, skin and bones, even internal organs. In 1998 the product "Apligraf," made of artificially grown skin, was confirmed as an engineered part of a body, which has since been regularly produced in tissue engineering centers to heal skin corrosions or burns. Today, researchers have not yet been able to produce functional organs or tissues, such as muscles, but the technology gives quite palpable promises that we will soon be able to get useful engineered organs or even to re-grow limbs. When engineering and cell cultivation in the laboratory are conducted for the purposes of transplantation, the technology is called regenerative medicine. Tissue engineering is a technology of in vitro tissue manipulation, which nowadays mainly uses stem cells in artificially created support systems that are set up for the execution of specific biological functions, particularly for the repair or replacement of parts of the tissue (like skin, cartilage, bone).

Eugene Thacker, one of the first humanists to discuss tissue engineering, points to the new comprehension of the body introduced with it: "Tissue engineering is able to produce a vision of the regenerative body, a body always potentially in excess of itself."[43] According to Thacker, because of the idea of regeneration the economy of body parts (transplantations, xeno-transplantations) has been replaced by the economy of auto-regeneration (regeneration of tissues from one's own cells), which is cyclic and proliferative (productive of a great number of cells with their division).[44]

Thacker introduces the concept of the regenerative body with reference to the function of proliferation. This is the immanent function of many types of cells, such as skin, cartilage, bone and muscle cells. Tissue engineering as a branch of biotechnology has relied exactly on these functions, which are already present in the body, but of which one can make good use of with in vitro manipulation before applying it back to the body. That is how regenerative medicine has gained its regenerative effect. In the next phase tissue engineering additionally relied on the even nobler functions brought about with stem cells. With their potentiality to become a particular cell, at which point the decision about differentiation depends upon need, stem cells shift the notion of regeneration to another, higher level. It is particularly stem cells which execute the auto-regeneration of a tissue within the body. The function of auto-regeneration has not only been acknowledged, but has

---

[43]Eugene Thacker, "The Thickness of Tissue Engineering: Biopolitics, Biotech, and the Regenerative Body," in: Stocker and Shöpf (eds.), *Ars Electronica 99*, p. 183.
[44]Ibid., p. 182.

been intensified with support of technology, in a manner that suggests great medical achievements, resulting in the optimization and prolongation of life.

A stem cell is a non-differentiated cell that has the ability of self-regeneration. During the process of reproduction, two daughter cells are created: the first one is identical to the original, but the other one is partially differentiated and more specialized. Somatic stem cells are located throughout the whole adult body, while embryonic stem cells are found only in the embryo. Stem cells enable several new ways of treatment. Today, the expression "advanced therapy" is well-established in EU medical legislation (EU Act [ES] number 1394/2007 of the European parliament and board), which divides advanced therapies into gene therapy, somatic cell therapy and tissue engineering. Advanced therapy uses principles of self-regeneration in tissue injury as well as in the treatment of cancer.

A stem cell's intrinsic capacity for self-renewal makes it the hallmark of the auto-regenerative capacity of the body, which could be determined as the fundamental distinction of living organisms in comparison to mechanic body simulations (but not their digital simulations as well). There are several hypotheses accounting for the capacity of self-renewal,[45] of which the most favored is the asymmetric hypothesis stating that one of the daughter cells keeps the peculiar stemness property of stem cells while the other is committed to differentiation. The immortal strand hypothesis (Cairns 1975) suggests that following DNA replication the oldest template strands are continuously retained by one of the newly generated cells (which will become a stem cell), while the other daughter cell will inherit all of the younger new template strands (this one will be committed to differentiation). This hypothesis can account for the origin of cancer cells, i.e. cancer stem cells, since the replication-induced mutations are all inherited by only one daughter cell, the non-stem cell.[46] The hub-niche hypothesis states that stem cells are maintained in a very specialized anatomical compartment where both proximal (cell surface molecules) and distal (secreted molecules) signals provide a definite microenvironment (the niche) controlling cell proliferation and differentiation, thus protecting the stem cell from exhaustion.

Regenerative capacity has been explored in nearly all tissues, and several factors have proven to play a role in auto-regenerative processes,[47] in which *proliferation* and *differentiation* are the fundamental ones assuring auto-regeneration. The cells that make up the early stages of embryonic development (blastocysts) are *pluripotent*, meaning they can differentiate into the three embryonic layers (the endoderm, the mesoderm and the ectoderm) that will create all the cell types that make up various tissues and organs. *Multipotent* and *unipotent* stem cells are those stem cells that have a limited differentiation ability and are found in the later stages of fetal and adult development, while the single-cell embryo (the zygote, a union of sperm and

---

[45]Monti and Redi, pp. 133–134.

[46]Riccardo Fodde et al., the Migrating Cancer Stem Cells consortium. www.mcscs.eu.

[47]Yufang Shi et al., "Mesenchymal Stem Cells: a New Strategy for Immunosuppression and Tissue Repair," in: *Cell Research*, Vol. 20, No. 5, 2010, pp. 510–518.

egg cell) is *totipotent*—it will form a new organism (the adult human organism consists of about a million billion cells). Mesenchymal stem cells, residing in every tissue and having the ability to self-renew and differentiate into almost any functional cell type,[48] were considered key players in auto-regeneration.[49] Mesenchymal stem cells are multipotent stromal cells, i.e. cells of connective tissue, which can be of a different quality depending on their age and location origin, both of which determine their regenerative capacity,[50] i.e. differential and proliferation potential. For example, umbilical cord derived mesenchymal stem cells are of a high quality and adipose tissue derived mesenchymal stem cells are of a low quality. In the body, adult stem cells referred to as mesenchymal stem cells have the ability to migrate toward tumors, injured and hypoxic sites.[51] There they can differentiate into any cell type depending on stimuli from the tissue.[52] Therefore, mesenchymal stem cells have become appraised as internal control factors regulating the auto-regenerative process, and there is a new belief that tissue stem cell regenerative power could surpass the need for external medical interventions during some auto-regenerative processes.

Because of the increase of sedentary life-styles and the aging of population, there is a socioeconomic interest invested in research and body engineering that promote the body's immanent capacity to self-regenerate. Options that are opened up by regenerative medicine promise the solution to several health problems (degenerative illnesses, cancer etc.), the transformation of the body itself, and improvements in the quality of life, even "rejuvenation," which actually means the prolongation of the life and active age of a social subject. Although the length of life and active age have both been prolonging since the human species began to make improvements in the quality of life (decrease of life and illness threats, variegation of food, etc.) and to conduct medical interventions, the options now made available by the regenerative body and enabled by biotechnology are displacing the limit of life beyond traditionally understood ones. This is significantly facilitated by working "from within" (or better, *with* the body itself) instead of manipulating the body and life "from outside," as was performed before with the help of mechanical or chemical

---

[48]Cristina Trento and Francesco Dazzi, "Mesenchymal Stem Cells and Innate Tolerance: Biology and Clinical Applications," in: *Swiss Med Weekly*, 2010, 140: w13121; Helena Motaln, Cristian Schichor and Tamara T. Lah, "Human Mesenchymal Stem Cells and Their Use in Cell-Based Therapies," in: *Cancer*, Vol. 116, Nr. 11, 2010, pp. 2519–2530.

[49]Maria G. Valorani et al., "Hypoxia Increases Sca-1/CD44 Co-Expression in Murine Mesenchymal Stem Cells and Enhances their Adipogenic Differentiation Potential," in: *Cell Tissue Research*, Vol. 116, Nr. 11, 2010, pp. 111–120.

[50]Carmen K. Rebelatto et al., "Dissimilar Differentiation of Mesenchymal Stem Cells from Bone Marrow, Umbilical Cord Blood, and Adipose Tissue," in: *Experimental Biology and Medicine*, July 2008; 233 (7), pp. 901–913.

[51]Yanique Rattigan et al., "Interleukin 6 Mediated Recruitment of Mesenchymal Stem Cells to the Hypoxic Tumor Milieu," in: *Experimental Cell Research*, 2010, Vol. 316, Nr. 20, pp. 3417–3424.

[52]Trento and Dazzi, "Mesenchymal Stem Cells and Innate Tolerance: Biology and Clinical Applications"; Motaln, Schichor and Lah, "Human Mesenchymal Stem Cells and Their Use in Cell-Based Therapies."

interventions. Today, the prospects of various branches of biotechnology are merging into the paradigm of a regenerative body. The next-generation therapies promised by synthetic biology perhaps originated with the idea of mechanical and chemical intervention (we should not forget that the objective of tissue engineering is to support organ transplantations). However, both therapeutic methods proposed by synthetic biology so far—to use the body's own proteins and to build molecular computers to enforce proliferation and suppress apoptosis—are actually contributing to the paradigm of a regenerative body by fostering the body's immanent life-affirmative functions and suppressing its life-taking functions.

It could also be claimed that the function of stem cells in the organism attests to a very important function in the body, which is *vitalization*. Accordingly, in acknowledging this function of stem cells, the recognition of the process of mortification in the body with the onset of illness (by Bichat) is getting supplemented with the recognition of an opposing process, that of *"vivification."* This is the process that testifies to *life as that which opposes death* as noticed by Foucault. The process of vivification with stem cells defies the natural process of mortification in the organism by assuring a constant resistance to the threats of illnesses and thus death. This issue could be linked to the notion of immunity as the ability to preserve and protect life, which is at the forefront of the contemporary debate on biopolitics. For nearly two thousand years, immunity has served almost exclusively political and juridical ends (a legal concept invented in ancient Rome), and only in the 1880s and 1890s did biomedicine acknowledge a new vital function: "immunity-as-defense."[53] Immunity is one of the central concerns of the contemporary philosophy of biopolitics, especially when political issues concerning society, defense and security are discussed.[54] Haraway warns about the constructiveness of bodies and their boundaries that occurs in the discourse of immunology. Applying Simone de Beauvoir's saying that one is not born but made a woman to the discourses of immunology, she notes: "one is not born an organism. Organisms are made."[55] With the biotechnological support of the body's internal force and strength, the discourse about immunity has only been intensifying.

Regenerative medicine as an intervention technology optimizing the body is in the midst of introducing a new vision of the body—a self-correcting,

---

[53]Ed Cohen, *A Body Worth Defending. Immunity, Biopolitics, and the Apotheosis of the Modern Body* (Durham, London: Duke University Press, 2009).

[54]This is being performed particularly following Foucault and his lectures on security, territory, and population at the Collège de France in 1977–78. Recently the issue was discussed in regard to September 11 by Jacques Derrida in: Jacques Derrida, "Autoimmunitiy: Real and Symbolic Suicides," in: Giovanna Borradori (ed.), *Philosophy in a Time of Terror: Dialogues with Jürgen Habermas and Jacques Derrida* (Chicago: University of Chicago Press, 2003), pp. 85–136. In regard to synthetic biology, the issue of security was discussed by Paul Rabinow and Gayamon Bennett in: Paul Rabinow and Gaymon Bennett, *Designing Human Practices. An Experiment with Synthetic Biology* (Chicago, London: The University of Chicago Press, 2012), particularly pp. 154–157.

[55]Haraway, "The Biopolitics of Postmodern Bodies. Determinations of Self in Immune System Discourse," p. 374.

self-improving, self-regenerating body. Foucault already analyzed some level of the technological intervention into life (combining the regulatory technology of life with the disciplinary technology of the body) and examined the example of the death of Franco, who was kept alive after he died, to present how the two systems of power coincided: that of sovereignty over death and that of the regularization of life. He noted: "And thanks to a power that is not simply scientific prowess, but the actual exercise of the political biopower established in the eighteenth century, we have become so good at keeping people alive that we've succeeded in keeping them alive when, in biological terms, they should have been dead long ago."[56] Today, however, the technological possibilities of regulating life and disciplining the body reach far beyond those available at his time. Thanks to the attainments of regenerative medicine, the power to make live now exceeds the limits of the "natural" life and body, much more so than was enabled by the institutionalization of medicine. The biological concepts of life and body need to be transposed, insofar as they are now both significantly traversed by technology. Ultimately, the idea of the regeneration of the body is generating a utopian vision of an immortal active life and body enabled by a reinforced constant process of vitalization to triumphantly defeat the natural process of mortification.

With this transformation of medical knowledge and technical possibilities, a shift is very slowly taking place that is related to the one described above. It is a shift from a mechanical paradigm to that of the (self-)regenerative body. The function of self-regeneration ought to be recognized as the essential function of the body and life, which itself already opposes the Cartesian notion of the objective body and thus modern medical thought, but which also has its especially explicit functional derivation in the potential of advanced therapy with stem cells. Modern medicine and biology operated with a concept of an organism as being composed of mechanical parts—this human body was thus understood as a kind of complex machinery—and this corresponded to the Cartesian causal comprehension of the body in the detection of local defects and the offer of direct pointed treatment. In accordance with this, medical treatment was conducted on the basis of the elimination or exchange of the damaged parts. In aesthetic surgery, the body was transformed mechanically as well, with direct plastic interventions into the body and insertions of artificial materials. Recent acknowledgements demonstrate that such methods are obsolete because the healthy parts of an organism are damaged collaterally. Advanced therapy on the contrary suggests the use of the body's own matter, which is to be implanted to improve the quality of the body's immanent ability to regenerate itself. Advanced therapy thus no longer suggests mechanical or chemical repair of the body, but instead develops options of stimulating the self-regenerative body. It is no longer appropriate to speak of *anatomo-clinical medicine* of the sort that Foucault was referring to in eighteenth- and especially nineteenth-century medicine. Since the second half of the twentieth century, medicine has been significantly altered by its biotechnological supports. Combining

---

[56]Foucault, *"Society Must Be Defended,"* p. 248.

biology with technology, biotechnology has been established as a techno-science or techno-knowledge. In our era, when engineering is highly advocated, even medicine itself, supported by biotechnology, has become a branch of engineering.

Let us finally discuss what this function of regeneration in research and medical practice actually means, as well as what the current prospects of regenerative medicine and the regenerative body in fact are. The vast range of prospects varies from simple techniques already being applied in practice, as in stem cell repopulation, to great ambitions, as in the case of efforts to help the body re-grow a whole limb or organ. Stem cell repopulation is a relatively simple technique of collecting stem cells from one body source (usually from adipose tissue) and their repopulation in another, injured part of the body (stem cells in breast reconstruction), where regeneration is then actually performed by the body itself. More complex approaches propose intervention by means of tissue engineering. We have already noted that tissue engineering is filling the void where native physiology or artificial implantable materials cannot sufficiently replace or repair the damaged tissues. Some tissues like bone or skin can effectively repair a small injury in a sufficient amount of time, while many tissues such as myocardium and cartilage do not regenerate properly without intervention. In the body, from early developmental phases onwards, embryonic stem cells produce their own extracellular scaffolds by secreting many types of molecules in the surrounding space, according to well-defined program of differentiation.[57] A great variety of natural scaffolds develops according to the spatial organization of the secreted molecules. Cells continue to proliferate and to organize themselves in order to build tissues and accomplish all of their natural functions. Researchers are now aiming to understand cell differentiation and functions by understanding cell-to-cell and cell-to-extracellular-matrix communication mechanisms. Cells receive signals from the environment to carry out their differentiation and proliferation programs through which they then form tissues or organs. The whole process of generating tissues, organs or limbs is actually very complex one, and regenerative medicine is on a quest to comprehend it in full, while "in the meantime" it is already able to operate on the basis of its partial knowledge with technologies that have already been developed according to it. Engineers strive to develop such matrixes that will enable in vitro engineering of functional tissue formations,[58] however up to this moment the most functional matrix has been proven to be the natural one (a decellularized organ). Because of cell division, cells can be multiplied in vitro, but a group of keratinocytes in a petri dish is not skin yet. Skin is a very complex organ. Skin or cartilage cell cultures that have been proliferated in a laboratory, for example, can be used to repair injured tissue when applied to the body. But in this case, they will get organized into a tissue formation by the signals

---

[57]Alfonso Barbarisi and Francesco Rosso, "Regenerative Medicine: Current and Potential Applications," in: Barbarisi (ed.), *Biotechnology in Surgery*, p. 75.

[58]Exogenous extracellular matrix for tissue engineering could be synthetic or biological. Gelatin, collagen, alginate, or synthetic polymers are utilized to constitute scaffold materials. Naturally occurring materials are composed of polypeptides, polysaccharides, nucleic acids, hydroxyapatites, or their composites.

coming from the body, meaning that the regeneration process will actually be completed or mainly performed by the body itself. However, it is another issue altogether to engineer a real functional tissue or even an organ completely in vitro, as for example a muscle with a contractile function and electrophysiological properties. It has been demonstrated that cardiac myocytes from neonatal rats—when mixed with collagen type I and matrix factors, cast in circular molds, and subjected to phasic mechanical stretch—reconstitute ring-shaped engineered heart tissue that display important hallmarks of differentiated myocardium. Data represent highly differentiated cardiac tissue constructs with action potential as the engineered construct was able to autonomously repeat mechanical stretching.[59] Tissue engineering has thus been delivering promises in heart valve tissue-engineering.[60] As regards real medical applications, in November 2008 a young mother in Spain was given a trachea transplant grown using her body's own stem cells. As regards the imminent prospects: in June 2010 the first transplant of bioengineered lung tissue into a living rat was performed at Yale University—the tissue performed its gas-exchange functions for 2 h after transplantation. In July 2010 lung tissue grown in culture functioned for 6 h when transplanted into living rats at Harvard University Medical School. At present a promising technique is organ (e.g. heart) regeneration involving the implantation of resorbable organic matrix (decellularized heart organ) with cells (e.g. cardio-myocytes), which could be differentiated from stem cells. With mechanic or electric stimulation, the attachment and alignment of the seeded cells

---

[59]Wolfram-Hubertus Zimmermann et al., "Tissue Engineering of a Differentiated Cardiac Muscle Construct," *Circ. Res.*, February 8, 2002; 90(2): 223–230.

[60]Tissue-engineered heart valves have been proposed to be the ultimate solution for treating valvular heart disease. Rather than replacing a diseased or defective native valve with a mechanical or animal tissue-derived artificial valve, a tissue-engineered valve would be a living organ, able to respond to growth and physiological forces in the same way that the native aortic valve does. Two main approaches have been attempted over the past fifteen years: regeneration and repopulation. Regeneration involves the implantation of a resorbable matrix that is expected to remodel in vivo and yield a functional valve composed of the cells and connective tissue proteins of the patient. A synthesis of a completely autologous fibrin-based heart valve structure using the principles of tissue engineering has already been demonstrated. See Tom Flanagan et al., "A Collagen-Glycosaminoglycan Co-Culture Model for Heart Valve Tissue Engineering Applications," *Biomaterials* Vol. 27, Issue 10 (2006): 2233–2246. Molded fibrin-based tissue-engineered heart valves seeded with ovine carotid artery-derived cells were subjected to twelve days of mechanical conditioning in a bioreactor system. Dynamic conditioning increased cell attachment/alignment and the expression of *á*-smooth muscle actin, while enhancing the deposition of ECM proteins, including types I and III collagen, fibronectin, laminin and chondroitin sulphate. This experiment from 2007 demonstrates that the application of low-pressure conditions and increasing pulsatile flow not only enhances seeded cell attachment and alignment within fibrin-based heart valves, but dramatically changes the manner in which these cells generate ECM proteins and remodel the valve matrix. Optimized dynamic conditioning, therefore, might accelerate the maturation of surgically feasible and implantable autologous fibrin-based tissue-engineered heart valves. See Tom Flanagan et al., "The In Vitro Development of Autologous Fibrin-Based Tissue-Engineered Heart Valves through Optimised Dynamic Conditioning," *Biomaterials* Vol. 28, Issue 23 (2007): 3388–3397.

within the matrix is enhanced.[61] The degree of tissue formation and differentiation of cells in vitro (e.g. into cardiac myocytes), as well as the contractile function and electrophysiological properties, are the factors upon which will depend the suitability of such tissues or organs for possible in vivo application.

That is to say, tissue engineering able to surpass current transplantation problems is our tomorrow. But perhaps even the bold ambition to re-grow the whole organs or limbs is not so remote. We know that the salamander has the ability to re-grow a whole limb if cut. It is a challenge to comprehend this function, but the interesting thing is that it is actually not foreign to humankind. During fetus development, this function is in full operation, while the whole body is immersed in a wet milieu similar to that of salamander. The human liver can re-grow after damage. The function of re-growth is related to the process of scarring, which protects the body from its dissolution into the environment and the invasion of the world (the other as discussed in the previous chapter) into the identity of the body. Re-growing and scarring stand in opposition, they are dialectical identities. There have been reports that a cut finger has successfully re-grown with matrix treatment (Lee Spievack 2008). As reported, it was important that the body was preserved opened, wounded during the process of "healing." Such acknowledgements force us to rethink issues of identity, body, growth, healing, life etc., which are at least within the Western context generally comprehended as self-contained entities and not according to their structural relations. Such is also the prospective of the regenerative body, a body able to re-grow its own parts and treat its own diseases, a body in a constant process of rejuvenation. In the already overpopulated world, however, what is driving this quest for regenerative body?

Foucault analyzed the emerging institutionalization of medicine in the context of normalizing society, when power took possession of life or at least took life under its care in the course of nineteenth century, at a time when "medicine becomes a political intervention-technique with specific power-effects. Medicine is a power-knowledge that can be applied to both the body and the population, both the organism and biological processes, and it will therefore have both disciplinary effects and regulatory effects."[62] The role medicine gained for biopower has only been intensified with the emergence of biotechnology, which instantly became the supporting technology of Foucault's so-called anatomo-politics of the human body, or what we prefer to call in this case the regenerative-politics of the human body. Regenerative medicine in particular is focused on the performances of the body, in optimizing its capabilities, extorting its forces, increasing its utilities. Additionally, biotechnology has become the supporting technology of the biopolitics of the population. Regenerative medicine manages life processes, particularly by improving the level of health, life expectancy and longevity. Regenerative medicine is therefore to be acknowledged as one of the leading technologies of contemporary biopower. The political role of regenerative medicine is crucial in slowing down the

---

[61]See *Initiation*: www.polona-tratnik.si/initiation.html, 7-21-2012.

[62]Foucault, *"Society Must Be Defended,"* p. 252.

process of aging, assuring quality of life, active aging, and instant regeneration. Last but not least, all of these mottos are represented in popular culture. The cultural tendency towards youth and the need to form one's own aesthetics of the body according to existing cultural standards in order to exhibit the healthy and fit condition of the body is growing, and in this regard regenerative medicine offers novel options and promises far-reaching solutions for sustainable corrections to the body. Regenerative medicine certainly contributes not only to the politics of the body but also to the politics of life. Regenerative medicine, supporting biomedicine, significantly consolidates the power to make live, established within the emerging normalizing society analyzed by Foucault. At present the power to make live testifies even more than ever to the fact that life and death are not natural or immediate phenomena that fall outside the field of power, but rather are decisively subjected to the mechanisms, techniques, and technologies of power.

Bibliography

G. Agamben, H. Sacer, *Sovereign Power and Bare Life* (Stanford University Press, Stanford, 1998)

L.B. Alberti, *On Painting*, trans. J.R. Spencer (Yale University Press, New Haven, 1970)

P.R. Anstey, Experimental versus speculative natural philosophy, in *The Science of Nature in the Seventeenth Century. Patterns of Change in Early Modern Natural Philosophy, Studies in History and Philosophy of Science*, vol. 19, eds. by P.R. Anstey, J.A. Schuster (Springer, Dordrecht, 2005), pp. 215–242

P.R. Anstey, J.A. Schuster (eds.), *The Science of Nature in the Seventeenth Century. Patterns of Change in Early Modern Natural Philosophy, Studies in History and Philosophy of Science*, vol. 19 (Springer, Dordrecht, 2005)

Aristotle, *The Physics*, trans. P.H. Wicksteed, F.M. Cornford (Harvard University Press, William Heinemann Ltd., Cambridge, MA and London, 1963)

Aristotle's Poetics, trans. L. Golden (Florida State University Press, Tallahassee, 1981)

ATR Intelligent Robotics and Communication Laboratories, "Geminoid HI-1," in *Human Nature. Ars Electronica 2009*, eds. by G. Stocker, C. Schöpf (Hathe Cantz, Ostfildern, 2009), pp. 218–221

J.L. Austin, *How to Do Things with Words* (Oxford University Press, London, 1962)

F. Bacon, *The New Organon* (Cambridge University Press, Cambridge, 2000)

M. Bakaoukas, The good life. An ancient Greek perspective. http://ancienthistory.about.com/library/bl/uc_bakaoukas4a.htm. 29 Dec 2009

M. Barasch, *Icon. Studies in the History of an Idea* (New York University Press, New York and London, 1995)

A. Barbarisi (ed.), *Biotechnology in Surgery. Updates in Surgery* (Springer, Milan, 2011)

A. Barbarisi, F. Rosso, Regenerative medicine: current and potential applications, in *Biotechnology in Surgery. Updates in Surgery*, ed. by Alfonso Barbarisi (Springer, Milan, 2011), pp. 75–94

R. Barthes, Rhetoric of the image, in *Image-Music-Text* (New York: Hill & Wang, 1964), pp. 32–51

C. Bernard, *An Introduction to the Study of Experimental Medicine*, trans. A.M. Henry Copley Greene (Henry Schuman, Inc., New York, 1949)

X. Bichat, *Physiological Researches Upon Life and Death* (Smith & Maxwell, Philadelphia, 1809)

E. Booth, *A Subtle and Mysterious Machine. The Medical World of Walter Charleton (1619–1707) Studies in History and Philosophy of Science*, vol. 18 (Springer, Dordrecht, 2005)

P. Bourdieu, *On Television* (The New Press, New York, 1998)

T. Campbell, *Bíos*, Immunity, Life, in *Bíos. Biopolitics and Philosophy*, ed. by R. Esposito (University of Minnesota Press, London, Minneapolis, 2008), pp. vii–xlii

G. Canguilhem, *The Normal and the Pathological*, trans. C.R. Fawcett (D. Reidel Publishing Company, Dordrecht, 1978)

G. Canguilhem, *Vital Rationalist*, trans. A. Goldhammer (Zone Books, New York, 2000)

© The Author(s) 2017
P. Tratnik, *Conquest of Body*, SpringerBriefs in Philosophy,
DOI 10.1007/978-3-319-57324-3

G. Canguilhem, *Knowledge of Life*, trans. S. Geroulanos, D. Ginsburg (Fordham University Press, New York, 2008)

E. Cohen, *A Body Worth Defending. Immunity, Biopolitics, and the Apotheosis of the Modern Body* (Duke University Press, Durham, London, 2009)

L. da Vinci, *Treatise on Painting*, trans. J. Francis Rigaud (George Bell & Sons, London, 1877)

A. Damasio, *Descartes' Error: Emotion, Reason, and the Human Brain* (G.P. Putnam, New York, 1994)

A.C. Danto, *The Philosophical Disenfranchisement of Art* (Columbia University Press, New York, 1986)

F. de Saussure, *Course in General Linguistics* (Philosophical Library, New York, 1959)

G. Deleuze, F. Guattari, *A Thousand Plateaus. Capitalism and Schizophrenia* (Continuum, London, New York, 2005)

J. Derrida, *Limited Inc* (Northwestern University Press, Evanston, IL, 1988)

J. Derrida, Autoimmunitiy: real and symbolic suicides, in *Philosophy in a Time of Terror: Dialogues with Jürgen Habermas and Jacques Derrida*, ed. by Giovanna Borradori (University of Chicago Press, Chicago, 2003), pp. 85–136

R. Descartes, *Meditations on First Philosophy. With Selections from the Objections and Replies*, trans. and ed. by J. Cottingham (Cambridge University Press, Cambridge, 2002)

M.-L. Dolezal, M. Mavroudi, Theodore Hyrtakenos' *Description of the Garden of St. Anna* and the Ekphrasis of Gardens, in *Byzantine Garden Culture*, eds. A. Littlewood, H. Maguire, J. Wolschke-Bulmahn (Dumbarton Oaks, Washington, D.C., 2002), pp. 105–158

R. Esposito, *Bíos. Biopolitics and Philosophy* (University of Minnesota Press, London, Minneapolis, 2008)

J. Fiske, *Television Culture* (Routledge, London, New York, 1987)

T. Flanagan et al., A collagen-glycosaminoglycan co-culture model for heart valve tissue engineering applications. Biomaterials **27**(10), 2233–2246 (2006)

T. Flanagan et al., The in vitro development of autologous fibrin-based tissue-engineered heart valves through optimised dynamic conditioning. Biomaterials **28**(23), 3388–3397 (2007)

V. Flusser, On Discovery, in *Artforum*, New York, vol. 26, no. 1 (September 1987)

V. Flusser, On Discovery, in *Artforum*, New York, vol. 26, no. 2 (October 1987)

V. Flusser, On Discovery, in *Artforum*, New York, vol. 27, no. 10 (Summer 1988)

V. Flusser, On Discovery, in *Artforum*, New York, vol. 27, no. 2 (October 1988)

V. Flusser, On Discovery, in *Artforum*, New York, vol. 27, no. 7 (March 1988)

V. Flusser, *Towards a Philosophy of Photography* (Reaktion Books, London, 2000)

V. Flusser, *Writings* (University of Minnesota Press, Minneapolis, London, 2002)

V. Flusser, *Into the Universe of Technical Images* (University of Minnesota Press, Minneapolis, London, 2011)

M. Foucault, *The History of Sexuality. Volume I: An Introduction* (Pantheon Books, New York, 1978)

M. Foucault, *The Birth of the Clinic. An Archaeology of Medical Perception* (Routledge, London, New York, 2003)

M. Foucault, *"Society Must Be Defended". Lectures at the Collège de France, 1975–76* (Picador, New York, 2003)

M. Foucault, *The Order of Things* (Routledge, London, New York, 2005)

M. Foucault, *"The Birth of Biopolitics". Lectures at the Collège de France, 1978–79* (Palgrave Macmillan Ltd., New York, 2008)

M. Foucault, *"Security, Territory, Population". Lectures at the Collège de France, 1977–78* (Picador, New York, 2009)

S. Franklin, M. Lock (eds.), *Remaking Life & Death. Toward an Anthropology of the Biosciences* (James Currey Ltd, Oxford, 2003)

H.G. Gadamer, *Truth and Method* (Continuum, London, New York, 2006)

O. Gal, R. Chen-Morris, Empiricism without the senses: how the instrument replaced the eye, in *The Body as Object and Instrument of Knowledge. Embodied Empiricism in Early Modern Science, Studies in History and Philosophy of Science*, vol. 25 eds. by C.T. Wolfe, O. Gal (Springer, Dordrecht, 2010), pp. 121–147

K.E. Gilbert, H. Kuhn, *A History of Esthetics* (Indiana University Press, Bloomington, 1954)

H. Gottweis, Genetic Engineering, Scientific-Industrial Revolution and Democratic Imagination, in *Ars Electronica 99. Life Sciences*, eds. by G. Stocker, C. Shöpf (Ars Electornica, Festival for Art, Technology and Society, Linz, 1999), pp. 122–134

S.S. Hall, *Mapping the Next Millennium. How Computer Driven Cartography is Revolutionizing the Face of Science* (Vintage Books, New York, 1993)

S. Halliwell, *Aristotle's Poetics* (Duckworth, London, 1986)

D. Haraway, The Biopolitics of Postmodern Bodies. Determinations of Self in Immune System Discourse, in *Knowledge, Power, and Practice: the Anthropology of Medicine and Everyday Life*, eds. by S. Lindenbaum, M. Lock (University of California Press, Berkeley, 1993), pp. 364–410

D. Haraway, The Cyborg Manifesto. Science, Technology and Socialist-Feminism in the Late Twentieth Century, in *The Cybercultures Reader*, eds. by D. Bell, B.M. Kennedy (Routledge, London, New York, 2000)

G.W.F. Hegel, *Science of Logic*, trans. George di Giovanni (Cambridge University Press, Cambridge, 2010)

M. Heidegger, *Basic Writings*, trans. D.F. Krell (Harper, San Francisco, 1977)

M. Heidegger, *Identity and Difference*, trans. J. Stambaugh (The University of Chicago Press, Chicago, 2002)

R. Hooke, *Micrographia*, Octavo (CD-Rom edition), 1998 (cop. The Warnock Library, London, 1665)

M. Jay, Scopic Regimes of Modernity, in *Vision and Visuality*, ed. by H. Foster (The New Press, New York, 1988), pp. 2–27

M. Jay, D. Eyes, *The Denigration of Vision in Twentieth-Century French Thought* (University of California Press, Los Angeles, Berkeley, 1993)

C. Klestinec, Practical experience in anatomy, in *The Body as Object and Instrument of Knowledge. Embodied Empiricism in Early Modern Science, Studies in History and Philosophy of Science*, vol. 25, eds. by C. T. Wolfe, O. Gal (Springer, Dordrecht, 2010), pp. 33–57

H. Landecker, On beginning and ending with apoptosis, in *Remaking Life & Death. Toward an Anthropology of the Biosciences*, ed. by S. Franklin, M. Lock (James Currey Ltd, Oxford, 2003), pp. 23–59

P. Laurie, *The Joy of Computers* (Hutchinson, London, 1983)

M. Lazzarato, From Biopower to Biopolitics, in *The Warwick Journal of Philosophy*, vol. 13, pp. 112–125. http://cms.gold.ac.uk/media/lazzarato_biopolitics.pdf. 27 June 2011

T. Lemke, *Biopolitics: An Advanced Introduction* (New York University Press, New York, London, 2011)

M. Lock, On making up the good-as-dead in a utilitarian world, in *Remaking Life & Death. Toward an Anthropology of the Biosciences*, ed. by S. Franklin, Margaret Lock (James Currey Ltd, Oxford, 2003), pp. 165–192

M. Lock, Twice dead: organ transplants and the reinvention of the death, in *The Body. A Reader*, ed. by M. Fraser, M. Greco (Routledge, Oxon, New York, 2005), pp. 262–266

M. Merleau-Ponty, Eye and mind, in *The Merleau-Ponty Aesthetics Reader. Philosophy and Painting*, eds. by G.A. Johnson (Northwestern University Press, Evanston, Illinois, 1993), pp. 121–149)

M. Merleau-Ponty, *Phenomenology of Perception* (Routledge, London, New York, 2005)

M. Monti, C.A. Redi, Stem Cells, in *Biotechnology in Surgery. Updates in Surgery*, eds. by A. Barbarisi (Springer, Milan, 2011), pp. 131–149

H. Motaln, Cristian Schichor, Tamara T. Lah, Human mesenchymal stem cells and their use in cell-based therapies. Cancer **116**(11), 2519–2530 (2010)

F. Nietzsche, *The Will to Power*, trans. W. Kaufmann and R.J. Hollingdale (Vintage Books, New York, 1968)

C.S. Peirce, Logic as semiotic: the theory of signs, in *Philosophical Writings of Peirce* (Dover Publications, Inc., New York, 1955), pp. 98–119

E. Pennisi, Synthetic genome brings new life to bacterium, in *Science* 21 May 2010: vol. 328, no. 5981, pp. 958–959. http://www.sciencemag.org/content/328/5981/958.full. 5 July 2012

Plato, *Republic*, trans. C.D.C. Reeve, (Hackett Publishing Company, Inc., Indianapolis, Cambridge, 2004)

P. Rabinow, G. Bennett, *Designing Human Pracitces. An Experiment with Synthetic Biology* (The University of Chicago Press, Chicago, London, 2012)

Y. Rattigan et al., Interleukin 6 mediated recruitment of mesenchymal stem cells to the hypoxic tumor milieu. Exp. Cell Res. **316**(20), 3417–3424 (2010)

C.K. Rebelatto et al., Dissimilar differentiation of mesenchymal stem cells from bone marrow, umbilical cord blood, and adipose tissue. Exp. Biol. Med. **233**(7), 901–913 (2008)

M. Riordan, *The Hunting of the Quark* (Simon & Schuster, New York, 1987)

R. Rorty, *Philosophy and the Mirror of Nature* (Princeton, Princeton University Press, 1979)

Y. Shi et al., Mesenchymal stem cells: a new strategy for immunosuppression and tissue repair. Cell Res. **20**(5), 510–518 (2010)

S. Sontag, *On Photography* (Picador, New York, 1977)

E. Thacker, The thickness of tissue engineering: biopolitics, biotech, and the regenerative body, in *Ars Electronica 99. Life Sciences*, eds. by G. Stocker, C. Shöpf (Ars Electornica, Festival for Art, Technology and Society, Linz, 1999), pp. 180–187

T. Todorov, *The Conquest of America*, trans. Richard Howard (Harper & Row, Publishers, Inc., New York, 1984)

C. Trento, F. Dazzi, Mesenchymal stem cells and innate tolerance: biology and clinical applications. Swiss Med. Weekly **140**, w13121 (2010)

M.G. Valorani et al., Hypoxia increases Sca-1/CD44 co-expression in murine mesenchymal stem cells and enhances their adipogenic differentiation potential. Cell Tissue Res. **116**(11), 111–120 (2010)

A. Vesalius, *De Humani Corporis Fabrica* (1543), available at: National Library of Medicine. http://archive.nlm.nih.gov/proj/ttp/flash/vesalius/vesalius.html. 1 June 2012

C.T. Wolfe, O. Gal (eds.), *The Body as Object and Instrument of Knowledge. Embodied Empiricism in Early Modern Science, Studies in History and Philosophy of Science*, vol. 25 (Springer, Dordrecht, 2010)

P. Woodruff, Aristotle on *Mimēsis*, in *Artistotle's Poetics*, ed. by A.O. Rorty (Princeton University Press, Princeton, 1992), pp. 73–95

S. Zielinski, *Entwerfen und Entbergen. Aspekte einer Genealogie der Projektion* (Walther König Verlag, Köln, 2010)

W.-H. Zimmermann et al., Tissue engineering of a differentiated cardiac muscle construct. Circ. Res. **90**(2), 223–230 (2002)

Index

© The Author(s) 2017
P. Tratnik, *Conquest of Body*, SpringerBriefs in Philosophy,
DOI 10.1007/978-3-319-57324-3